"If you think work, life, and world economies can continue without intervention, you're not alone. But the reality is that nothing is permanent. Everything is either disrupted by outside factors or improved from within. King and Petty help us understand the role we play in the latter."

—Brian Solis
Digital Anthropologist, Futurist, Bestselling Author

"For the past 17 years my research has centred on the question of 'how do we increase our capacity to adapt to a world of accelerated change?' This question, unlike a closed puzzle where we know what the answer will look like before we begin, or an open puzzle where we need to figure out the answer as we go, this question behaves more like a mystery that keeps us on the edge of our seats and induces our curiosity and fascination every step of the way, and just when we think we know the answer, another mystery emerges. *The Rise of Technosocialism* is a brilliant maze of mysteries and will inspire you to decide which road to take to create the social, cultural, economic and technological future you desire."

—Rocky Scopelliti
Futurologist and Author of *Youthquake 4.0* and *Australia 2030*

"Brett and Richard do a fantastic job of making us think about the impact of technology on the future of humanity. If you are looking for a book that will make you reflect on humanity's purpose, challenge preconceived ideas around economics, and show you a different perspective, you will love *Technosocialism*."

—Henri Arslanian
Bestselling Author of *The Future of Finance*
and Adjunct Professor at the University of Hong Kong

"Brett King and Richard Petty's view of the future is grounded in reality. This is not just futurists talking about flying cars, in this case they deliver practical preparation for a new world that we're just entering. The two authors aggregate the major trends facing our world and put them into a realistic and readable account of how technology is going to change our world moving forward. From vaccines to digital currency all of these factors are going to play into a future where the fabric of society itself is transformed by digital technology."

—Richard Turrin
Bestselling Author of *Cashless: China's Digital Currency Revolution*

BRETT KING & DR. RICHARD PETTY

THE RISE OF

TECHNOSOCIALISM

How Inequality, AI and Climate will Usher in a New World

FOREWORD BY

DR. HARRY KLOOR

CEO and Executive Founder, Beyond Imagination

www.technosocialism.com

Marshall Cavendish
Business

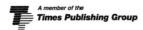

To my great friends Clayton Fitts and Stephen Phillips
(aka Living Legend), who are always there when I need them
— Brett King

For JLA
— Richard Petty

CONTENTS

FOREWORD

What flavour is your Technosocialism?

My best friend Peter Diamandis often credits me with coining the phrase, *technosocialism*. I don't know if that's true, but I started using that term around 1988 at the inaugural class of the International Space University (ISU). This was a magical time for me with 100 plus graduate students from around the world. The Mainland China and Soviet students would talk in private about the poor conditions and the lack of opportunity created by communism. Soviet students expressed their excitement at the prospect of change, especially come mid-summer when their government canceled all history tests, admitting that everyone had been taught a fiction about their countries past.

On paper, the Soviet Union's communist/socialist constitution provided for equality and prosperity for all, but in reality, their systems lead to the greatest atrocities the world had ever seen. Those at the top not only had all the power, but all the wealth.

The students from Europe advocated for socialism, a lighter version of communism where, in theory, you'd get to own your property, but many, myself included, thought socialism was plagued by the same problems of communism, with the elite pulling the strings and public benefits never being equal. As I see it, people are not ideal machines who'll work to the best of their ability and happily share the harvest of their labour with everyone equally.

Let's face it—driven people work their minds and/or bodies as hard as they can. Some are altruistic, others lack motivation and will take advantage of flaws in any political system to manipulate and worm their way to the top. Unfortunately, it doesn't take many of these bad apples to ruin a system. Humans can be noble in reason, but they can also be deeply flawed.

What if humans were removed from the equation? What if you fulfilled individual needs through technology? It stands to reason that exponential technologies like AI and robotics could and would eventually lead to a different form of socialism, one where all your needs could be met by innovation. What happens if housing, food, medicine, energy, education, and transportation are of high quality and nearly free due to innovation? You get what I call, Technosocialism.

Now my flavour of Technosocialism does not involve the government, rather, the innovations and businesses that fill those needs are created by smart entrepreneurs. We are living in the early stages of this system already. For instance, the internet and search engines like Google and DuckDuckGo have democratized and demonetized knowledge. Cellular technology and smartphones now provide 65% of the globe with low-cost, high quality voice and video communication at a fraction of the cost that it used to be. AI, combined with the Internet of Things, is rapidly evolving education. Soon a full education via an online AI teacher will be available for everyone for a few bucks a week. Education will be tailored to each student's unique abilities and interests—far better than what we now have.

The coming of humanoid and other robots powered by AI is feared by many because it will eliminate most, if not all labour by humans. Now, in full disclosure, I run a company called Beyond Imagination (BE), which was co-founded by myself, Ray Kurzweil, Paul Jacobs, Tony Robbins and other luminaries. This year we completed construction of Beomni, a highly advanced, general purpose humanoid robot with an AI engine which is trained via a human pilot. Initially, Beomni will take jobs away from robots by removing the need for employers to develop an autonomous robotic AI driven workforce due to rising wages. The breakthrough innovation is that Beomni is piloted by a human while in Avatar Mode and wherever there is a Beomni, people around the world with a cellular or WiFi connection will be able to access it and work remotely. Beomni will democratize opportunities for all and create a world of instant services which will apply not only to hospitality, construction, mining, and agriculture, but to white collar jobs such as engineers, nurses, and doctors.

Initially, our system will create jobs—not eliminate them, but over time, the system will learn tasks via the human operation, evolving to becoming semi-autonomous, enabling one person to work in many places at once. Eventually, our systems will evolve to become fully autonomous which will eliminate jobs, but this is a good thing. Beomni, along with other autonomous robots, will increase prosperity for all, global poverty will decrease, as will the cost of everything, from medicine to transportation. As robots like Beomni become ubiquitous, many other needs of mankind will, in turn, be democratized and demonetized.

Automated technical labour means being able to see a doctor anytime, anywhere. The ability to see a robotic doctor with an AI that can actually provide preventative care, will lead to maximum health for everyone, especially in conjunction with the Internet of Things, which will provide your AI doc with the ability to see health progress as well as problems and act immediately. Coupled with Quantum Computing, AI will rapidly advance and as it does, this will lead to amazing technological improvements in renewable energy and battery storage. Cheap, clean energy coupled with evolving robots that can perform all that labour will make the cost of food, other material goods, and infrastructure like housing and roads nearly free.

Now, some versions of Technosocialism argue that, as technology replaces jobs, there'll be a need for universal funds so that basic needs can be met. In my opinion, this is flawed reasoning that will lead to far more problems than it will cure. But don't simply agree with me, instead think for yourself and explore the possibilities. That's why I like this book you're about to read—*The Rise of Technsocialism* will open up discussion on a range of topics and potential future roads. You don't need to agree, but you do need to think and debate about them.

Discourse may broaden your perspective on Technosocialism—you may realize that historically, at every stage of technological development, as innovation replaced certain jobs, more jobs were created. Automation in farming and manufacturing led to more jobs, fewer injuries and deaths, and greater output. Yes, there will be people who find themselves out of work, and we need to make sure there is a safety net and retraining for able working people. But the whole point of Technosocialism is to provide for the needs of everyone while removing government interference.

If the cost of our needs is drastically reduced while the quality of these services and goods is dramatically improved, the number of hours we all need to work is reduced. More free time can then be spent on family, exercise, entertainment and loftier goals. Until the robots completely take over most labour, people can receive an income to help others, help the environment, and care for the elderly.

Until COVID hit, I took two weeks off every year to teach middle school kids that they were "the exponential generation" . . . that they are growing up into a world of AI and robot servants. I taught about the Internet of Things, quantum computing, 3D printing, and autonomous cars. They quickly grasped that the simple labour jobs their parents have would not be available to them in the future, and that more than ever they needed to study hard to develop their intelligence, their creativity, their artistic as well as logical parts of their brains. Imagine if instead my lesson to them was, "Hey you're going get a UBI when you grow up, so don't worry about learning, the government will take care of you."

My flavour of Technosocialsm will take care of you, as long as you work a little bit. UBI already exists for the children and grandchildren of wealthy people, but according to David Kleinhandler, 90% of these grandkids will lose their family's fortunes because they lack an understanding of the value of work, have never struggled to achieve anything, and have no understanding of money. Creating a global economy of uninspired people sounds like a nightmare.

Lastly, I'm a big believer in technological solutions for the environment—that's why my friends Jon, Jeff, and I originated a Carbon Extraction Prize at the 2017 XPRIZE Visioneering event. Elon Musk funded that prize to the tune of over 100 million this year. It's not enough to stop damaging the environment, we need to repair it, and we can do so via Technosocialism. Imagine a million robots working to clean up every river and the ocean of all the waste mankind has dumped in it. That's my flavour, what's yours?

Dr. Harry Kloor
CEO and Executive Founder
Beyond Imagination

PREFACE

The 21st century is going to be the most disruptive, contentious period humanity has ever lived through. It will challenge our most sacred ideologies around politics, economics and social constructs. It will force humanity to adapt in ways we can't yet imagine.

There is much to be highly optimistic about, but it will require humanity to unite in respect to our collective goals and purpose. With the emergence of Artificial Intelligence (AI) we are on the verge of perhaps solving the biggest mysteries of the universe, but AI will also allow us to automate society to provide untold abundance and prosperity. We will soon have the technology to extend our lifespan, to make humanity a multi-planetary species, and to provide for the basic needs of every man, woman and child on the planet.

Within a decade or two, we will have retooled most of the world's energy systems to be completely renewable, and we are starting a journey to reimagine education, healthcare, housing, consumption, food and agriculture with economies built for the 21st century.

The best way to describe the likely outcomes, we believe, is as we have articulated in these pages. This will be the rise of a technology-driven collective social consciousness and purpose. If that term makes you think of a classical right-wing conservative view of socialism, of debates around Venezuela's economic collapse or the writings of Karl Marx, let us stop you right there. This is absolutely not what we are advocating.

We are simply looking at the fact that multiple trends, converging forces, and looming social issues will cause the entire world to challenge traditional views of functioning democracies, capitalism, and Western political ideals as rolling crises continue to impact the globe. It could best be described as a global social movement that forces huge change in respect to inclusion and policies by government, the private sector as well as non-governmental organisations. If we could find a better term to describe the evolving geopolitical and economic landscape, we'd embrace it. Neo-capitalism? Nope—capitalism is a core driver of those unintended consequences, creating social division and poor incentives. Twenty-first century democracy? No. This doesn't get close to describing the impact social media, artificial intelligence and technology have had—and will continue to have—on politics. Populism? No, populist movements are more likely symptoms of a failing system and—a reaction to globalization, not a solution to the political and social divide.

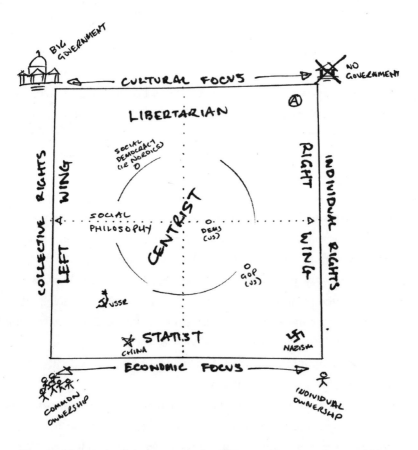

Figure 1: Historical political spectrum, last 200 years. (Source: author graphic)

Consider the broad political spectrum we've experienced during the 20th century. While in the United States the Democratic Party is often considered somewhere between centrist and radical left, the reality is that from a global historical perspective they're more centre-right than communist. Things like universal health care, free education and strong social security don't make governments historically far left; in fact, many democracies that field these basic services would be right of centre today.

There are a few major forces that are likely to turn the classic political spectrum on its head during the 21st century. Firstly, high levels of automation will shift "Big Government" culture back toward the centre as technology allows us to provide all the services we'd expect from a modern government, but at a fraction of the cost and effort we saw during the 20th century. Secondly, the effects of climate, ongoing pandemics and growing inequality will push global governance to focus increasingly on broad collective rights and action. Lastly, changes in value systems will shift priorities of communities away from classic capitalism to more sustainable and inclusive options. More on this in our opening chapters.

Cooperation between governments globally to combat climate change and rolling pandemics must create more collaborative governance. Corporations in this future will be forced to mitigate high-levels of automation that disrupt human labour with strategies that focus on their social commitment and environmental responsibility—otherwise their brands will be toast.

Economically, we are entering an era of explosive uncertainty. Over the last 40 years we have assembled a collection of the richest, most profitable individuals and companies the world has ever seen. But as change has accelerated we're leaving increasingly large swathes of society behind. The number of wealthy individuals or the size of corporate profits within a geographical footprint can no longer be considered a singularly positive measure of macro-economic success.

As technology reshapes supply and demand economics, capital markets will adapt and labour participation will be turned on its head. If your economy hasn't educated, retrained and supported your people sufficiently, along with investing in next-generation infrastructure and focused on 21st century competitiveness—the impact will be brutal. Technology-based displacement of traditional jobs is one thing, but the realization by the vast majority of your populace that they have no economic future, no real and valued stake in society, no share in the success they see others enjoying, is a more philosophical issue. All this, while new emerging and highly profitable industries suffer severe labour shortages due to lack of adequate planning, lack of access to education and misplaced immigration policies.

Technosocialism is not something entirely new, but it is a philosophy driven by the people, for the people, empowered by incredible advances in technology, and reinforced through constant challenges to the status quo. Both policy and technology will need to work together to ensure the collective basic needs of our communities, emphasizing greater societal cohesion and improved action against uncertainty and volatility. If socialism is characterized by the needs of the collective, and technology allows us to provide for those and at a much lower political and economic cost, then logically government and the universal common good will be much more efficient and economical.

In this book we look at a range of possible outcomes, but as a Futurist and an Entrepreneur and Academic we are most concerned about the adaptability of our social, political, and economic models to a future fast bearing down upon us.

History teaches us that this future is largely inevitable, and that we are, on the whole, woefully unprepared. Why? Because we've spent the last 200 years creating, empowering and incentivizing the machinery and the systems that have birthed uncertainty and inequality. Our short-term focus on GDP, jobs, quarterly results and election cycles accentuates our inability to properly plan more than a few years out and leads to us kicking the can down the road far too often. This myopic short-termism is only getting worse, with greater dysfunction to come if collectively we don't adapt to a new reality. Humans must shift to much longer-term planning cycles and broader economic participation if we are to survive as a species.

This book is about facing the harsh realities of our future. It is about understanding the evolution of the social movements we see today and how they will unfold, and having real grown-up policy conversations that mitigate risks to our stability, freedoms and a healthy future. We will also look at the possible outcomes if we simply double down on the flawed broken systems of today.

We hope *The Rise of Technosocialism* gets you thinking about your future, your family's future, and that of your community. As optimists, we hope you see that these changes don't have to divide us and that we can build a more prosperous and inclusive future for all.

But first we need to get on the same page.

Brett King   **Richard Petty**

The Futurist   Entrepreneur and Academic

ACKNOWLEDGEMENTS

As per usual, writing a book like this is a multi-year effort and requires a tribe in terms of support and delivery.

Particularly I wanted to thank the team at Marshall Cavendish and Times Publishing Group along with the global publishers who have once again got behind our book with enthusiasm and great gusto, particularly Melvin Neo, Janine Gamilla, Norjan Hussain and Michael Spilling. As always, I would be remiss not to thank the teams at Moven and Provoke for their outstanding support and understanding as weeks became months of disruption as the book went through rewrites due to COVID impact—particularly Richard Radice, Bryan Clagget, JP Nicols, Jason Henrichs, Cassie Leblanc, Kevin Hirshorn, Liesbeth Savereins, Carlo Navarro, Elena Liman, and Marek Forysiak. To Jay Kemp and Tanja Markovic, your ability to keep my schedule running, to keep the revenue coming and to look after my mental and physical health simultaneously, is truly inspirational and I'm always blessed to have you guys on the team.

To Richard Petty for sticking with this project when the pandemic threw our world and project into chaos, and for believing we can change the world.

Once again, I'd like to thank the coffee shops who made this book possible. The Algonquin Hotel Lobby Lounge in NYC, The Ludlow Hotel in the East Village, Starbucks Reserve (Icon Siam) and the Coffee Club (Riverside Plaza) in Bangkok specifically. Your WiFi and cappuccinos were the fuel that kept me writing.

But the greatest of thanks go to my family – Katie (Miss Metaverse), Charlize (Charlie) and Thomas (Mr T) in particular for being the gravity that keeps me centered. To Matt and Hannah who continue to make me proud. And to my dad who remains my greatest fan and PR agent.

There are many, many more, and you know who you are, that keep me engaged and sane daily. I love you all and I'm a better person for having you in my life.

Brett King (BK)

Writing a book is difficult at the best of times. Writing one during a pandemic when many of the really big things like technology, the economy, environmental issues, politics, and social interactions are changing more rapidly than ever before poses unique challenges. Writing *The Rise of Technosocialism* has taken us on a long journey; but one that, for us, has been very worthwhile.

What began as an idea discussed over coffee in New York several years ago, took shape over discussions held in several different countries as Brett and I fought to make time to discuss and debate in-person the issues that are important to us, and then to capture our thoughts on paper. Brett and I have been friends for decades and we often have thought to do a project like *The Rise of Technosocialism* together. I am so pleased that he had the inspiration for *The Rise of Technosocialism* and insisted that this was the one we should do.

We have been well supported by the team at Marshall Cavendish and Times Publishing Group along with our global publishers. Particular thanks to Melvin Neo, Janine Gamilla, Norjan Hussain, and Michael Spilling.

My thanks to the family and friends who stand with me during times good and bad. My habitual reclusiveness during periods of workaholism and stress—alas, those periods are more frequent and sustained than is healthy—may create the false impression that I take you for granted; I assure you that this is never the case. Special thanks to my sister, a gifted writer who possesses compassion and strength in equal measure and is a model of inherent goodness.

To those who inspire me: students past and present who taught me to listen, teachers who encouraged me to think critically rather than to be critical, and the women and men of science who should be celebrated and recognised to a much greater extent than they are (your time surely will come, for without you all is lost).

Richard Petty

CHAPTER I

EXPLOSIVE UNCERTAINTY

> "The purpose of government is to enable the people of a nation to live in safety and happiness. Government exists for the interests of the governed, not for the governors."
> —Thomas Jefferson

It took just 21 days to bring the world's largest economy to its knees. From 20th January 2020 when the first COVID-19[1] patient tested positive in Seattle, until 11th February. That period saw the start of the US stock market's worst bear run in history—one that wiped out more than a third of the market's value before it was done. All this from a virus that measures just 0.125 micron, or 125 nanometers, in diameter. In comparison, a single human hair is 400x larger than the SARS-CoV-2 "hidden enemy".

By the end of May 2020, one in four Americans had filed for unemployment benefits, putting the total unemployed in excess of 40 million persons. Prior to the coronavirus crisis America had never recorded a single week of one million jobless claims, but by the 2020 US election we'd averaged one million new jobless each week for six months (peaking above two million claims in the second-last week of May). The International Labour Organization (ILO) estimated that by the end of Q2, at least 195 million jobs globally would have been wiped out, along with 6.7% of total working hours. But these negative effects were not evenly distributed across economic classes.

In the United Kingdom, COVID-19 deaths in the most deprived areas were more than double those in the most affluent neighbourhoods (Office for National Statistics). The US' Federal Reserve chairman at the time, Jerome Powell, said in a speech in May that 40% of households earning less than $40,000 per annum lost at least one job in the household during March 2020. In a study based on predicted job and wage losses, the Aspen Institute Financial Security Program and the COVID-19 Eviction Defense Project concluded that 19 to 23 million renters in the United States were at risk of eviction through the end of 2020, representing up to 21% of renter households. Similarly, Amherst Capital, a real estate investment firm, estimated in June 2020 that 28 million households (64 million people) were at risk of eviction due to COVID-19. As late as November 2020, 88% of New York restaurants had failed to pay their rent. Even with the vaccine rollout, these economic impacts will be felt for much longer.

The NYC Health Department announced during the crisis that African-American and Latino people were more than twice as likely to be killed by the virus than whites living in the city. This is not a statement on genetics, but on healthcare disparity between the poorest and more affluent segments of society.

Failures in the market

It could be argued that COVID-19 wasn't so much a failure of medical science, as a failure of the free market and governance. All things being equal, the US could have easily afforded to have enough ventilators, antivirals and other medicines stockpiled for an imminent pandemic—but they didn't. The free market simply couldn't respond fast enough. A functioning healthcare market can't materialize in the space of weeks for a disease that appeared out of nowhere that impacts millions simultaneously. The US healthcare system is argued as an example of a free market model, but COVID-19 showed it wasn't a fair and equitable market for all Americans facing a disease that doesn't check your bank balance before infecting you. It showed that the free market does not inherently have core metrics for the greater public or social good—not when the stock market was skyrocketing while daily deaths climbed to surpass the number of deaths of 9/11, with more than 100,000 businesses shuttered and at least 30 million people relying on unemployment benefits and stimulus payments to survive.

When we've been warned about possible pandemics for more than 20 years, it is also hard to argue that the pandemic was a failure of imagination. In 2005 the US Department of Health and Human Services, along with the CDC, produced an Influenza Pandemic Response Plan[2] that anticipated exactly the type of pandemic scenario we found in COVID-19. We didn't know it would be COVID-19, but we knew pandemics were coming. Why? Because in human history they're simply a regular feature.

> *"Of course, the thing that people ask: 'What keeps you most up at night in the biodefense world?' Pandemic flu, of course. I think everyone in this room probably shares that concern."*
> **—Health and Human Services Secretary Alex Azar at the National BioDefense Summit (17 April 2018)[3]**

The WHO had also been response planning for a Spanish Flu-type influenza pandemic since at least 2004 (see WHO Pandemic Preparedness[4]). When push came to shove, implementing this plan became fraught with international politics, debates on the science, flawed communications, and poor coordination between city, state, federal agencies, countries, nation states and multilateral organizations. Governments that were well prepared and acted immediately and decisively weren't immune to either the virus' effects or the economic fallout either. This is not meant to be a political statement. The reality is that it has been 100 years since the last major pandemic, and despite all that time to prepare, this disease still threw us into global chaos. As this book goes to press, we're still years away from understanding the widespread effects of COVID-19 and the societal changes that will be necessary to return to some sense of normalcy.

Bill Gates famously spoke at TED in 2015 outlining the potential of a pandemic like the 2020 coronavirus, encouraging us all to work towards an effective global response capability.

When COVID-19 hit, people thought his predictions so uncanny that conspiracy theorists suggested he had created the virus to show people he was right, and to profit from vaccine production. Imagine being Bill and Melinda Gates, spending billions of dollars to reduce poverty globally and treating diseases like polio with success, only to be accused of doing all of that so they could plant microchips somehow embedded in future COVID-19 vaccines to control your brain? The fact is, Gates wasn't especially prescient, he knew—as did the entire global collective of immunologists and epidemiologists—that a pandemic was simply just a matter of time.

Debates raged about the efficacy of lockdown, about Sweden's atypical approach and why Asian nations had fared so much better than countries like the US. Protesters around the world marched to encourage release from lockdown. Medical personnel in major cities were already at their wits' end, pushed to their limits physically and emotionally. As the winter of 2020–21 descended upon the US, spikes in cases led to a worsening of the pandemic.

COVID-19 illustrates the potential for failures in our political, social and economic systems. What will happen when we face even worse crises?

Early responses

Actions taken by some governments did show clear benefits in slowing the impact of coronavirus early in the pandemic, although these strategies were often rejected by other governments. In Taiwan, arrivals from Wuhan were subject to health screenings even before human-to-human transmission was confirmed. By 1st February 2020, Hong Kong, Taiwan and Singapore had all implemented travel restrictions on passengers arriving from mainland China, even when the WHO initially maintained (in error) that such restrictions were unnecessary.

Following SARS in 2003, Taiwan established a central command centre for epidemics. By 20th January 2020, it was coordinating the government's response to the coronavirus. The command centre compiled a list of 124 "action items," including border controls, school and work policies, public communication plans and resource assessments of hospitals[5].

On 20 January, the Taiwan CDC announced that the government had under its control a stockpile of 44 million surgical masks, 1.9 million N95 masks and 1100 negative-pressure isolation rooms. Taiwan's vice president at the time happened to be a prominent epidemiologist. He gave regular briefings showing people when to wear a mask, the importance of hand washing, and explaining that hoarding masks prevented frontline health workers from accessing the equipment they needed. It seems like a simple, straightforward approach now, but was uncommon in the early days. As of the first half of 2021, Taiwan had just 187 deaths.

In Singapore everyone that presented with flu- or cold-like symptoms or a fever were immediately tested for coronavirus. The government ran full-page ads in the local newspapers, and TV and radio commercials urging people to stay home if they were sick. Back in 2003, as a result of SARS, Singapore built a task force across multiple government agencies to coordinate interventions and messaging during any future pandemics. This task force was tested in 2009 during the H1N1 pandemic and again in 2016 during the Zika outbreak. It was reassembled in January 2020 for SARS-CoV-2. Singapore established more than 1,000 testing clinics by mid-February across the city state. New York City's Manhattan Island is a tenth of the size of Singapore, but by June 2020 they had just 100 testing sites—in mid-April they only had nine. Singapore suffered 33 deaths.

Hong Kong was a city deeply affected by the SARS outbreak of 2003. Both of us lived in Hong Kong throughout SARS and saw the transformation firsthand. Small dark alleyway's full of discarded trash were cleaned up. Every citizen wore a mask when out in public. Every port, every bank, every major shopping mall and building had thermal imaging technology and temperature sensors deployed—if you had a fever, you were sent home or sent to hospital. When you got home you took off your shoes outside your residence. Once inside, you washed your hands before touching anything else. You changed your clothes and washed them. The Hong Kong government issued a stay-at-home order on 28th January. New York didn't do so until 20th March, the UK on 23rd March. Hong Kong had just 200 deaths.

This level of discipline and compliance in the face of coronavirus was generally unseen in the West. In Italy, mayors of affected cities had to threaten to visit homes hosting graduation parties with flamethrowers to break up gatherings. In the US the states of Kansas, Michigan, North Carolina, Ohio, and Florida, all saw street protests against stay-at-home orders, with many people refusing restrictions, and certainly refusing to mask up. At Miami Beach, Cancun, and New Orleans, partygoers celebrated Spring Break and Mardi Gras. Brady Sluder, a 22-year-old student from Milford, Ohio proclaimed: "If I get corona, I get corona... At the end of the day, I'm not going to let it stop me from partying. I've been waiting, we've been waiting for Miami spring break for a while. About two months we've had this trip planned, two, three months, and we're just out here having a good time."[6]

Within two weeks of Spring break, hundreds of students around the country who attended such events were infected with coronavirus; one week later, more than a dozen had died. In Louisiana, where local authorities thought the risk of infection was low, a month after Mardi Gras there were 6,000 confirmed cases in Orleans County alone, and nearly 400 deaths. President Trump continued to push for political rallies, and it has been suggested he may have been responsible for around 30,000 infections and 700 deaths.[7] The London School of Hygiene and Tropical Medicine pulled together a list of superspreader events around the world and found that about 20% of infected people are responsible for 80% of coronavirus cases globally, because of events where social distancing and mask-wearing were ignored. Still today you will find an abundance of Americans who claim masks don't work and the shutdown of the economy was an infringement on their rights, or even illegal. It didn't help that the CDC had sent out mixed messages earlier in the pandemic due to medical PPE shortages.

In December 2020, the US accounted for just 4% of the world's population, but over 25% of the world's COVID-19 cases.[8] Somehow those numbers just seem unbelievable—how could the world's most advanced economy get its response so fundamentally wrong?

Laws in place in Europe would make apps like Singapore's TraceTogether illegal. In much of the West, individual freedoms have increasingly been emphasized over the collective needs of society, and when coronavirus hit, our freedom to act against the interests of our fellow humans was indeed tested. The act of wearing a mask became a simple test—are your individual rights (the right to refuse a mask) more important than those around you (who you might infect)?

Gideon Lichfield, *MIT Technology Review*'s editor-in-chief, put it this way: "We are so tightly interconnected that a virus can reach each one of us, yet so insular that we cannot conceive of what happens in one place repeating itself in another." When presented with clear evidence that some countries were handling the crisis better than others, the results were largely dismissed by those under siege.

There were some unintended side effects from the shutdown that drew attention to other systemic issues confronting humanity.

Within just a few weeks of a virtual global shutdown of some of the world's largest cities pollution levels had plummeted. European cities like Milan, Rome, Barcelona and Paris all recorded drops of around 50% in nitrogen dioxide levels.

In the Indian state of Punjab, for the first time in 30 years, residents were able to see the Himalaya Mountains more than 100 kilometers (62 miles) away. In the city of Venice, canals were so clear that you could see fish swimming on the canal bottom—something that was last observed almost a century ago. Endangered turtles in Brazil were seen laying eggs on beaches normally overrun by people.

Globally, it's been 100 years since the last global pandemic, namely, the H1N1 pandemic of 1918–19, often called the Spanish Flu. In that time we've debated future pandemics, we've prepared—we've spent billions of dollars preparing—and yet when COVID-19 hit we abandoned all that preparation. Why?

Overwhelmingly, coronavirus has accentuated economic uncertainty at a time when historically we are already at one of the most economically contentious periods in human history. With the richest economies humanity has ever seen, and at a time of incredible technological

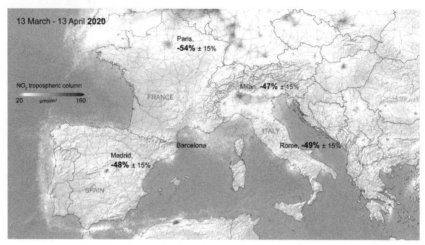

Figure 1: Nitrogen dioxide levels in Europe during coronavirus containment.
(Image Credit: KNMI/ESA Copernicus)

advancements, why can't we fix a problem like the coronavirus? Is it getting better or getting worse?

As a species you'd think that a threat to our fellow humans would crystallize support for global action to combat the spread of a virus, to eliminate poverty or reduce the impact of climate change on the planet. As yet, we just haven't got our act together to mitigate those risks. In fact, in many cases we can't even agree on whether these issues are real, let alone set meaningful policies that will result in remedial action.

We debate reasonable carbon reductions and emissions goals for our economies, when a minimum 60-centimetre (two-foot) rise in sea levels and a 2°C temperature rise is already largely baked in for 2050, even if we reduce emissions to zero immediately.

We've known for 100 years that humans were damaging the environment and that it would reduce air and water quality, along with potentially much more significant impacts. Seven to eight million people die annually from pollution, more than twice the deaths caused by COVID-19, and yet we trade off human lives for fossil fuel profits. As the evidence mounts that this damage is going to permanently change our coastlines, our agricultural industries, and will result in mass migration and food scarcity, we still debate the possibility of a global species-wide response. Why?

A failure of imagination?

Today 1.6 billion people lack adequate housing. In the US, since COVID-19, it is estimated that the number of homeless has climbed to over 2.5 million people, while more than 17 million homes in the US lie vacant. More than 10% of the world's population go hungry, and anxiety regarding food security has gone ballistic during coronavirus. During the best of times the US destroys 40% of its annual food production, more than 63 million tons, while in the US alone 38 million people go hungry annually. Around the world, panic buying at grocery stores and supermarkets ended up inflating the problem of food waste as people found that they couldn't eat all the food they had hoarded at the start of the coronavirus crisis. And having all that excess toilet paper didn't really help that much either.

The crisis that has unfolded around the world over the past couple of years gives us an important opportunity for self-examination and for the rethinking of systems that have failed us and will likely do so again. Will we seek to simply return to normal, or are we prepared to carve out a "new normal" that works for the future of all humans?

After the deaths of George Floyd, Ahmed Arbery, and Breonna Taylor, tens of thousands of protestors took to the streets in US cities. Twenty-five cities across 16 states[9] imposed curfews as those protests turned into riots. For young African American men living in the United States today, the sixth leading cause of death is police violence.[10] These protests across the US set police, National Guard, secret service and informal private militias against a collection of angry, disappointed and disillusioned citizens pleading with the government to finally address systemic racism and injustice—a reasonable social discourse. But protests over injustice and inequality have been steadily increasing in frequency for years now, even before this backlash against police actions. As you'll learn in our later chapters, the frequency and size of global protests is increasing by orders of magnitude, illustrating a fundamental dysfunction in our societies.

During the last week of October 2019, the governments of Lebanon and Iraq resigned in the face of relentless protests. A week later, the government of Bolivia did the same. In the preceding 12 months, leaders in the US, United Kingdom, Chile, Hong Kong, France, Indonesia, Netherlands, Peru, Haiti, Syria, Israel and Russia all faced political protests, ranging from tens of thousands to more than a million individuals. On the 19th and 20th of September 2019, six million protestors from across 185 countries took place in the largest global protest of its kind—protesting the lack of action against climate change. On 6th January 2021, approximately 30,000[11] Trump supporters stormed the US Capitol Building and its surrounding areas, demanding that the US election result be overturned.

Looked at in aggregate, you could make a strong case that modern democracy and capitalism is failing large cross-sections of humanity and all the other species we share the planet with.

Globally, coronavirus has impacted poorer citizens much harder than the affluent. With shutdowns, loss of income and very little access to the booming stock market, the perception of two different economic realities is one that is hard to avoid.

The virus accentuates inequality

Today the majority of the planet faces economic and social uncertainty at a time when humanity as a whole has never been wealthier and more technologically advanced. Statistically, by most measures this is the single best epoch in human history to be alive, with the lowest levels of poverty, famine, infant mortality, and disease, along with increases in longevity, affluence, and education.

Paradoxically for the poorest inhabitants of the wealthiest democracies, it is as if they have been thrust into a sort of neo-Middle Ages, where feudal landowners and the political elite have robbed them of the economic potential once imagined. Coronavirus has amplified these effects with poorer and middle-class citizens having been impacted harder by the virus.

The disparity between the so-called 1% "rich" elites and the remaining 99% is greatest in wealthy democracies like the United States and the United Kingdom. In few places is this skewed wealth distribution more visible than in Silicon Valley itself. Median income in California hasn't changed in 25 years, but in that same period housing prices have climbed

187%. A recent study from the workplace chat app Blind showed that 70% of tech workers earning a six-figure income still couldn't afford to buy a house or apartment in the San Francisco Bay area.

Ironically, the San Francisco property market was hit very hard by COVID-19, but not due to the economic recession. The change in working from home policies from players like Google, Facebook, Twitter and Apple led to a significant percentage of tech workers seeking accommodations outside of the Bay Area, changing a demand trend that was more than two decades in the making.[12]

History teaches us that this level of inequality does not go unanswered. Political and social movements aren't borne out of intellectual or political debate, they are birthed by social upheaval. When we look at the drivers behind the rise of Trump, Brexit, Boris Johnson, Bolsonaro and Le Pen, the perceived threat to long-held traditions or "culture" is often voiced in mainstream media, but the rate of technological change may be emerging as perhaps the most influential factor.

Released in 2019, the Edelman Trust Barometer Report[13] reported that 47% of people globally think that technological innovations are happening too quickly and will lead to changes that negatively impact "people like me". Fifty-nine percent of people believe they do not have the training and skills necessary to improve their employment prospects, and 55% think that automation and other innovations are already taking jobs away. The uncertainty factor that has already stimulated widespread dissent, protests and debate is snowballing.

At the heart of this book is our assessment that four primary stressors are converging to produce acute, long-term economic uncertainty that will continue to threaten social cohesion. These stressors will not only lead to a rethink of conventional economics and political policy, but a new renaissance for humanity. We know, that's a big statement.

Global resistance to the factors stimulating inequality, accentuated by rising concerns over technology-based unemployment and the mounting outcomes of climate change, will coalesce into a collective action that will attempt to force us out of the chaos. The pandemic is just a momentary crisis thrown into the mix, but more pandemics are likely too. This emerging global movement is already being led by a generation of people who reject conventional wisdom, who are ideologically geared toward more inclusive thinking, who possess greater social consciousness, and who believe that technology can be leveraged to fix the world's stickiest problems. They want to push us forward, while the old guard is frantic about us moving too fast, reminiscing about the good 'ole days. It's a recipe for multigenerational conflict for sure; but when politicians tend to be older and more nostalgic, this becomes a broader social conflict.

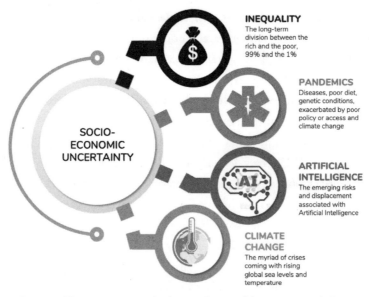

Figure 2: The major stressors leading to the era of disruptive social change.

Just like during the Industrial Revolution, politics and economics will need to evolve for a very different future. Neither a Western-style democracy founded on unbridled capitalism nor communism founded on Marxist principles will be able to restore cohesion in the face of these unstoppable forces. Socialism won't solve most of the stickiest problems either, but a more coherent social consciousness will absolutely be required. Technological advances can fix our stickiest challenges, but also can accentuate inequality and division.

Humanity has never faced this level of global uncertainty.

Back to the future

If we're going to get to a future where we are not immobilized by rolling global crises, where we see less political and social conflict, and where we see broader economic participation, we need to deal with the problem of inequality.

Inequality is measured at a national level using what is known as the Gini coefficient. The Gini coefficient is a number between 0 and 1; where 0 might represent a world where everyone makes the same income, and 1 where a single individual makes all the income (everyone else has zero). In the United States this tells a clear story—inequality today is the same as it was during the Great Depression of the 1930s, perhaps even worse, given purchasing power.

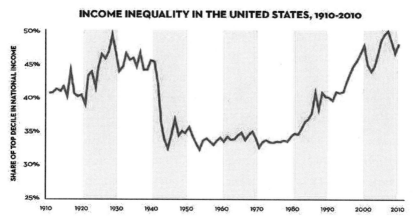

INCOME INEQUALITY IN THE UNITED STATES, 1910-2010

Figure 3: Gini coefficient in the US from 1910 till today. By the close of 2020 the estimated Gini coefficient was 0.54 (the highest in recorded history).

While many other Western nations have suffered from similar acceleration in inequality over the last few decades, the United States' form of extreme capitalism has its downsides. As economist Thomas Piketty noted in his book *Capital in the Twenty-First Century* the level of inequality presently in the United States is "probably higher than in any other society at any time in the past, anywhere in the world." And that was before the coronavirus pandemic accentuated the problem.

Authors Jonathan Tepper and Denise Hearn argue in their book, *The Myth of Capitalism: Monopolies and the Death of Competition*, that the failure of capitalism in the US is because America has gone from an open, highly-competitive marketplace to an economy where a few very powerful companies dominate key industries like technology, banking, pharmaceuticals and energy. This lack of competition has subsequently consolidated profits, reduced broader economic participation and created much of the imbalance we see today.

If left untouched, could capitalism right itself? Could capitalism solve these global problems of inequality, providing the stability society once again needs? History shows capitalism simply isn't incentivized to fix large-scale social problems. Capitalism's driving force is economic growth, not social policy.

Capitalism rewards companies and markets that produce economic returns. It does not reward them to deliver on social issues or dedicate parts of their business to the common good. In quarterly and annual results reporting, it's very rare to see an analyst hold a company's feet to the fire because they had chosen profits over the wider needs of society. Capitalism and stock markets simply don't have metrics that require companies to act ethically and in the best interest of the citizenry at large, unless those metrics are enshrined in law (usually after clear abuses). If it did, we wouldn't have lung cancer from cigarettes, pollution and carbon emissions from energy and fossil fuel corporations, obesity from poor-quality fast foods, medical bankruptcies due to healthcare costs, and so forth. The coronavirus crisis emphasized that healthcare systems long optimized for profitability were demonstrably fragile when faced with a global pandemic. A single serving president was able to render mute the decades of pandemic planning that the US had embarked on.

For those arguing that capitalism will fix these issues over time, need we remind you of the climate crisis we face because those issues weren't fixed 40–50 years ago by the market, or the simple inequity of providing COVID-19 tests for professional athletes attending sporting events when they were not made available for residents and staff in nursing homes? Climate aside, since the 1970s we've known of the effect that fossil fuel pollution has on the general populace. We could have easily accelerated green energy tech, but the market seemed much more willing to trade off profits against the inconvenient annual culling of city dwelling humans impacted by poor air quality.

Conventional political commentary suggests that capitalism is an integral part of an economic system that guarantees individual rights. But capitalism has failed to prevent the greatest crises we have faced, thus drawing attention to the fact that there needs to be greater balance between individual rights and the common good for an economy to function for all citizens long-term. In this way, climate change, inequality, and rolling pandemics over the next couple of decades will push economic policy towards the centre.

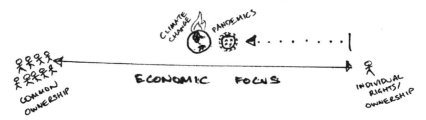

Figure 4: Economic uncertainty driven by inequality, pandemics and climate change will force capitalism to address common concerns over individual returns.
(Source: author's own illustration)

This is not so much about economic theory as about the application of economics for broader social goals. Individual rights can certainly be guaranteed, but only within a framework where they don't harm others. Democratic socialist economies like those in Scandinavia demonstrate that some balance is possible here.

Additionally, our Gen-Y and Gen-Z citizens are increasingly less concerned with asset ownership and accumulation, as they saw their parent's wealth decimated during the Global Financial Crisis and coronavirus pandemic. They are generations that may never be able to afford to buy their own home, especially in cities like Hong Kong, New York, London, Sydney or Tokyo. In its place, shared ownership structures, shared asset services, and the sharing economy mean that assets like homes and cars may fall out of fashion. Social media and an increasingly collective and tribal view of humanity have also led to broader policy engagement across the political map.

Figure 5: AI-based public resource management will allow big government to be low cost, reducing the arguments against social-leaning policy. (Source: author's own illustration)

Many conservatives argue that big government is inefficient, and that private enterprise and the free market can do better at resource allocation when it comes to economic growth. Underlying that idea is the notion that economic growth is fine regardless of the effect that it has on equity and access. AI-based government services and resource allocation will tip these old assumptions on their head. Big, broadly effective government mechanisms will become economically viable because automation will dramatically shrink government footprints and bureaucracy.

What is technosocialism?

In the pages of *Technosocialism*, we're definitely arguing for the reform of capitalism in the 21st century. Increasing levels of inequality in the world are likely to lead to an economic engine serving an ever-shrinking portion of society, along with risks that policy that is susceptible to lobbying efforts and vested interests will prolong this cycle indefinitely.

What is technosocialism? It isn't a political movement, it's a social outcome. Firstly, it resets long-term economic growth within a framework that does not harm the economy at large, while ensuring the maximum participation of all citizens in the economy. Secondly, it allows big government capability, with strong investment in technology infrastructure that radically improves government productivity—therefore eliminating most of the funding and budgetary objections that government programs would normally be subjected to.

If not technosocialism, what are the alternative outcomes for the planet over the next 50 years? We see four possible outcomes for today's society. They lie across two broad axes: collective vs individual and chaotic future vs ordered future as illustrated in Figure 6 on the next page.

- **Neo-Feudalism**: Unbridled capitalism that rejects the need for greater equality and fails to bring broad economic growth as employment and consumption slide. Long-term division between the rich elite and the poor reaches a melting point with rolling revolutionary actions and protests as the middle-class evaporates. The super-rich get access to longevity technology, AI, and abundance in closed enclaves, while mass unemployment, hunger, and disease are the norm for those outside.

- **Luddistan**: Rejection of technological advancements like AI. A slow and inept climate response stalls the global economy, while repeated crises contract population growth. Major coastal cities become uninhabitable due to rising sea levels. Food scarcity and hunger explode as crops fail.

- **Failedistan**: The largest economic states enter chaotic and reactive rule of law as the climate and markets collapse due to lack of planning, forethought and action. Global migration due to climate change is in the hundreds of millions. Borders collapse and resource wars rage. Governments collapse.

- **Technosocialism**: Society becomes highly automated, replacing most human labour. Technology advancements make housing, healthcare, education and basic services ubiquitous and low cost. Capitalism is re-engineered toward long-term sustainability, equality and the advancement of humanity as a whole. Climate mitigation efforts generate centuries-long global economic cooperation.

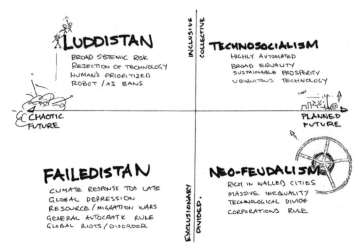

Figure 6: The likely possible futures humanity faces. (Source: author's own)

If you object to Luddistan or Failedistan as terms or classifications, you can think of them as simply Techno-prohibitive or Collective Failed States. You may feel that there are other alternatives to these scenarios, and we absolutely welcome that debate. The conclusions we've drawn are due to long historical analogies and studying human behaviours as we react to rolling crises. You'll have plenty of opportunity to debate with us as you continue to read.

As you read about technosocialism, we hope you will see this is not a political debate. This is more accurately a philosophical and economic policy debate. A philosophical argument in respect to the true purpose of humanity, the goals we strive toward as a species, and whether or not greater equality serves those goals and that purpose. A debate in respect to what sort of economic theory is needed to express those goals and promote inclusiveness as a core construct of our society, leading to broader happiness and greater prosperity. A debate, at its core, about what purpose the economy serves—is the economy's rightful role to empower a small segment of individuals in society? Or should it serve the needs of all citizens first and foremost?

Over the next 30 years humanity will face numerous crises that will accentuate the division between rich and poor, and expose the increasing failures of the free market to address the biggest problems that our planet has ever faced. We have choices as to how we deal with these issues. Those choices will determine the outcomes for humanity with varying degrees of success. Which choices will we make? Ones that benefit all humanity, or ones that benefit the few?

Endnotes

1   COVID-19 or SARS-CoV-2 is the class of coronavirus that impacted the world in 2020.

2   CDC National Pandemic Response Plan (2005 & 2017): https://www.cdc.gov/flu/pandemic-resources/national-strategy/index.html.

3   "Advancing Biodefense", Biodefense Summit Transcript, 2019: https://www.phe.gov/Preparedness/biodefense-strategy/Pages/advancing-biodefense-transcript.aspx.

4   Source: World Health Organization https://www.who.int/influenza/preparedness/pandemic/en/

5   See the *Journal of the American Medical Association*, "Response to COVID-19 in Taiwan Big Data Analytics, New Technology, and Proactive Testing", Wang et al, 3 March 2020.

6   CBS News, 18 March 2020.

7   Source: Stanford University Paper: https://papers.ssrn.com/sol3/papers.cfm?abstract_id=3722299.

8   Source: Fox News: https://www.fox5ny.com/news/us-has-4-of-the-worlds-population-but-more-than-25-of-global-coronavirus-cases.

9   Source: CNN.

10 Source: US News "Police Violence a Leading Cause of Death for Young Men", 5 August 2019. https://www.usnews.com/news/healthiest-communities/articles/2019-08-05/police-violence-a-leading-cause-of-death-for-young-men.

11 Source: NBC News quoting National Park Service's estimate: https://www.nbcnews.com/politics/congress/live-blog/electoral-college-certification-updates-n1252864/ncrd1252964#blogHeader.

12 CBS San Francisco: https://sanfrancisco.cbslocal.com/2020/10/10/covid-exodus-home-for-sale-listings-soar-as-pandemics-economic-impact-grips-san-francisco/.

13 See Edelman Trust Barometer Report: https://www.edelman.com/sites/g/files/aatuss191/files/2019-04/2019_Edelman_Trust_Barometer_Technology_Report.pdf.

THE PYRAMID OF INEQUALITY

On 24th January 1848, James W. Marshall found gold at Sutter's Mill in Coloma, California, resulting in the start of the California Gold Rush. Three decades later the gold rush had dissipated and employment was in rapid decline. Rather than blaming the declining mining industry and absence of gold for the decline in jobs, Chinese immigrants were blamed. In 1882 the Chinese Exclusion Act was signed into law by US President Chester Arthur, creating a 10-year moratorium on Chinese immigration. Historically we can see that Chinese immigration had little to do with the negative economic impact of the end of the gold rush. Politically, however, it became an easy excuse for the economic slowdown. It's a pattern we saw repeated in the 2016 US elections.

Research following the 2016 and 2020 US elections found that college-educated voters generally preferred Clinton or Biden, but the best predictor of whether someone would vote for Trump was usually correlated with anxiety around economics, education level, and the pace of change. If you were young and college educated, you tended to be more centrist or left leaning. If you were older and white or hadn't attended college, you were more likely to swing right.

As we watched the populist movement boom, individuals articulated their concerns around the rapid pace of change, new technologies, or immigration policies that might impact their prospects in the future. Looking at the world's largest economy, and perhaps most poignant populist experience, it is fair to say that Trump's rise to power was built on the unintended collapse of the "American Dream".

If you know your history, the term "American Dream" was coined by James Truslow Adams in his 1931 book *The Epic of America*. He described it as a "dream of a land in which life should be better and richer and fuller for everyone, with opportunity for each according to ability or achievement." The principle was simple: if you worked hard, sacrificed and took some risks, you could attain the sort of success that was never dreamed possible by your parents or forebears. The American Dream also guaranteed that your children would be better off. But this dream relied on continued economic growth, equitable distribution of wealth and continued opportunities for the next generation. Since the 1980s this economic potential has been dwindling for the vast majority of the US population.

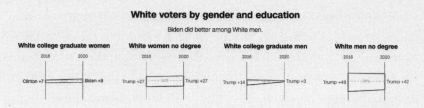

Figure 7: White non-college educated voters supported Trump in 2016/2020.
(Source: CNN Politics)

Ultimately, Trump's success in 2016 (and substantial support in 2020) came down to the fact that 60–64% of white working-class voters (those without college degrees and representing a third of American adults) supported him over Clinton[1] or Biden[2]. Four critical factors stood out as independent predictors of Trump's success on election night.

1. **Anxiety about cultural change**—concern over the "American way of life" being diluted through foreign influence.
2. **Immigration**—closely correlated with the above, but also related to concerns about employment.
3. **Economic uncertainty**—fear over immigrants and technology taking jobs and ever reducing wages, along with rising costs of living and housing.
4. **Failed education system and rising student loans**—61% of white working-class men believed college education to be risky and without merit. This correlates with ballooning college tuition fees and student debt in the US.

Practical economic effects weighed far heavier on the US population than climate change or concerns about ageing infrastructure. It came down to food on the table, a roof over the head, access to healthcare, and putting money in the bank. Those are problems that arguably shouldn't even exist in the US today. In future chapters we'll see how economic uncertainty erodes freedoms and happiness also.

During the 2016 and 2020 US Democratic Party primaries, the other surprise for many was the strong performance of Bernie Sanders. Sanders was unusual given he was essentially promoting what was seen as an anti-capitalist socialist agenda. The election of Alexandria Ocasio-Cortez in January of 2019 to the US House of Representatives, along with Bernie Sanders and Elizabeth Warren's continued success in the 2020 primaries, reinforced a debate around more inclusive policies in general. Indeed, Biden was forced to evolve some of his language and policy positions during the 2020 election as a result.

Ironically, the basic underlying fears supporting both the populist movement and socialism were very similar. Looking beyond the politics, the grassroots support at both ends of the political spectrum was coming from increasingly fearful constituents looking for radical solutions to perceived threats to their economic future.

"The rich get richer, the poor get the picture"[3]

While it is true that the French Revolution of 1789 was one of the first modern expressions of "power to the people" or *demos-kratos*, the Industrial Revolution that followed in the 1800s caused Western governments to deal with a sudden surge of popular social movements and big social policy gaps. Workers formed trade unions, giving them real political clout for the first time. The US, India, and others successfully challenged Britain's model of colonialism. The suffrage movement for women exploded out of New York and London, following a civil war in the US triggered by changing views on slavery. A social awakening for the world in respect to human rights was paralleling the revolution in economics and industry.

Economists looking back at the pre-Depression era in the United States attribute Henry Ford's production line as a key mechanism in allowing greater wealth distribution. The post World War II boom in manufacturing and technology innovation allowed the United States to create a robust middle class. Today, real wage growth has been conspicuously absent in the

US for most of the population for more than 40 years. The tendency to blame immigrants or external factors for this lack of improvement, rather than systemic policy bias and politics, is a well-worn political strategy reaching back to the California Gold Rush. Is there a more rational explanation for such a long-term erosion of the core middle class?

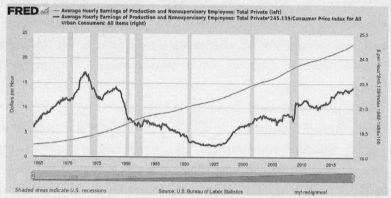

Figure 8: Real Wage growth in the United States today remains below 1970s levels (Source: BLS)

We debate minimum wage a great deal in the United States, but the reality is that if US real-wage growth had kept pace with the consumer price index and productivity growth, then the minimum wage would be more than $24 per hour today. This relatively flat wage growth is devastating for the economy on a long-term basis as it shrinks consumption, and it is at the heart of the growing inequality in the United States.

The Reagan years (1981–89) oversaw a period of policy and strategy that attacked big government spending and the power of unions. Both led to a decade long assault on mechanisms that had previously acted to slow income inequality. Bill Clinton contributed to this widening gap by opposing collective bargaining, deregulating the financial services industry, reducing welfare exposure and signing into law the NAFTA trade agreement. In 1987 the film *Wall Street* aired, giving us a glimpse into the persona of Gordon Gekko, an opportunistic and hungry Wall Street trader. Rather than pricking Wall Street's collective consciences over increasing inequality, Gekko became the poster child for traders and financiers the world over. Let's not even get started on the likes of Enron manipulating the deregulated Californian energy market.

In 2008 as the Global Financial Crisis emerged and the depth of the subprime mortgage portfolios unfolded, Michael Douglas, the actor who played Gekko, was asked at a United Nations event "whether he bore some responsibility for the behaviour of the greed merchants who had brought the world to its knees thanks to his encouragement.[4]" Douglas dodged the question, but the relationship between the "trader" mentality, the deregulation of the financial services industry through the Reagan and Clinton years, and rising income inequality is hard to ignore.

This systemic attack on the poor was also at work in the late 70s and 80s in the United Kingdom. The UK government faced off against trade unions as pressure on wages emerged. The government attempted to freeze wages and wage increases to slow inflation, but in doing so went against previous agreements with trade unions. The winter of 1978–79 became known as the "Winter of Discontent". Trash piled up on city streets, grave diggers went on strike, truck drivers refused to work overtime and gas stations were closed across the country as fuel deliveries slowed to a trickle. Train drivers and nurses followed. 22 January 1979 was the biggest individual day of strikes since the Great Depression, with many workers staying out indefinitely after the protests finished. More than one million workers were laid off as the debate continued to rage over the role of trade unions, inflation, and pricing.

Figure 9: Saatchi and Saatchi advertising campaign in 1978
bemoaning national strikes. (Source: Saatchi & Saatchi)

Margaret Thatcher ran on a platform of restricting trade union power, and on 3rd May 1979 she successfully won a general election. It wasn't until 1985 that she broke the back of the trade union movement, and even then, inflationary pressure didn't subside for some years. Thirty years later, wage growth in 2015 was at its lowest point since the 1860s. The UK is yet to even return to pre-financial crisis wage levels, and that was before both COVID-19 and Brexit have negative impacts on job and wage growth. Job growth had already collapsed and was on a downward trend after the Brexit decision, but GDP growth slumped 11.3% in 2020—the largest fall in economic output in 300 years.

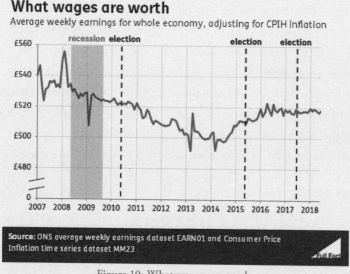

Figure 10: What wages are worth.

Looking at inflation, cost of living, and economic growth over the same period, declining real wage growth in the US and UK correlates with changes to legislation regarding trade unions and the deregulation of financial markets. The combination also resulted in a redistribution of wealth away from the middle class to the richest segments of society. Structural changes in both economies around technology and automation have also attacked traditional industry and employment patterns, accentuating the impact of shrinking wages. When you factor in sluggish wage growth, with massive increases in housing costs and inflation, real spending power for the middle class has been on the decline since the 1980s, and that's very bad for economies.

Trickle-down economics

This brings us to another policy element that could be argued to have stimulated growing inequality in the US over the last 40 years.

Reagan supported a hotly contested supply-side economic theory posited by the economist Arthur Laffer. These days when the "Laffer curve" is mentioned, most economists dismiss it outright. Laffer's theory was that by decreasing taxes, taxation revenue would actually increase as a result of greater economic activity—to be fair, he wasn't the originator of the idea, he just popularised it. This supply-side theory is otherwise known by the vernacular "trickle-down economics" today. In August 1981, Reagan succeeded in passing The Economic Recovery Tax Act, a policy based largely on the Laffer curve. In the four years that followed, federal tax revenue fell by an average of 13%. In 2019, Trump awarded Arthur Laffer the Presidential Medal of Freedom, and when the Republicans took control of both houses in 2017, they again cut taxes based on Laffer's arguments.

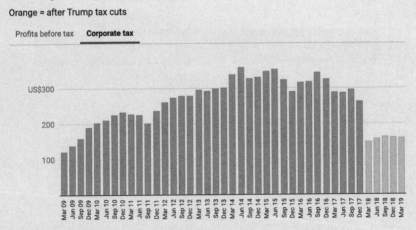

US corporate profits and tax

Orange = after Trump tax cuts

Profits before tax **Corporate tax**

US$300

200

100

Mar 09 Jun 09 Sep 09 Dec 09 Mar 10 Jun 10 Sep 10 Dec 10 Mar 11 Jun 11 Sep 11 Dec 11 Mar 12 Jun 12 Sep 12 Dec 12 Mar 13 Jun 13 Sep 13 Dec 13 Mar 14 Jun 14 Sep 14 Dec 14 Mar 15 Jun 15 Sep 15 Dec 15 Mar 16 Jun 16 Sep 16 Dec 16 Mar 17 Jun 17 Sep 17 Dec 17 Mar 18 Jun 18 Sep 18 Dec 18 Mar 19

Chart: Greg Jericho · Source: FRED · Get the data · Created with Datawrapper

Figure 11: Corporate profits.

The Economic Recovery Act of 1981 arguably represented the most significant reduction in US taxes in the history of the nation. But following that, Reagan's team introduced new taxes that hit the middle class materially. The Tax Equity and Fiscal Responsibility Act of 1982 (TEFRA) increased government revenue mainly by tightening up rules on depreciation, leasing, contract accounting and investment tax credits. The Social Security Amendments of 1983 sped up planned increases in payroll tax rates and, among other things, doubled the social security tax that lower- and middle-class working families paid. The Deficit Reduction Act of 1984 changed rules on interest exclusions, income averaging and such. The Omnibus Budget Reconciliation Act of 1987 closed a few loopholes and extended a telephone excise tax. And the Tax Reform Act of 1986, while it lowered the top individual income tax rate to 28% from above 50%, contained enough offsetting changes that, for the first two years after enactment, it raised tax revenues.

Effectively, between 1981 and 1989, Reagan doubled average income taxes for working-class taxpayers, while lowering taxation on the richest Americans from 74% percent to 28%. He created huge tax loopholes for corporations and incentivized companies to shift jobs offshore.

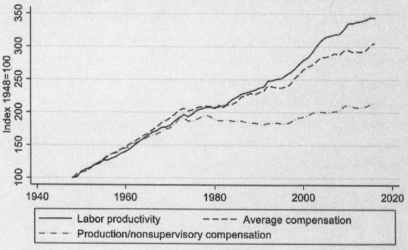

Figure 12: Real wages tracked labour productivity until the Reagan years.
(Source: BLS/BEA data)

From George Washington to Jimmy Carter, real wages tracked productivity growth in the United States. As companies became more productive and richer, they generally raised the wages of their workers commensurate with that productivity increase. This was a core factor in creating the strongest middle-class growth that the world had ever seen post World War II. Before Reagan, one working parent could support a middle-class family. After Reagan, with both parents working, the average American household was still worse off than in the preceding decades.

"Reaganomics", as it was known, fundamentally attacked wage equality, but it may not have been immediately evident at the time. Especially as the American public was sold on the biggest tax cuts in US history at the start of his first term in office.

Tax cuts for the rich accentuated wealth capture by the wealthiest Americans, leading to accelerated asset growth at the top of the pyramid from 1980 to today. Similar effects occurred in the UK, Australia, and elsewhere. Overall, the coronavirus pandemic seemed a boon for the wealthiest Americans. While nearly 40 million people filed for unemployment during the first months of the pandemic, the billionaire class in the US collectively saw their net worth increase by over $10 trillion in the same year[5].

What were the results of these macroeconomic changes?

They led to the Global Financial Crisis, brought Trump to power, enabled Brexit, kicked off the Occupy movement and Arab Spring, to name a few. While some news outlets shouted "Globalism and Immigration is to blame!" In reality, we've seen four decades of assault on real wages for the middle class, structural changes to the economy without real policy or strategy to address fundamental shifts in equality, and structural changes moving us away from traditional industry to knowledge workers, but without the commensurate investment in education and training to adapt the labour force. It strains credulity to argue that the fate of the US economy is down to a group of 10–12 million illegal or undocumented immigrants[6] who rank among the poorest segments of society, when the wellbeing of the middle-class itself has been under assault for 40 years.

The US has two very different outcomes when it comes to healthcare, depending on household wealth. The same is true for access to justice in the courtroom, access to reasonable treatment by the police force and public safety services, access to education, access to political representation and other social goods. While some in the US remain unaware of this ballooning inequity, the data doesn't lie. Today, the American Dream has given way to two Americas—one for the rich and one for the poor. The percentage of Americans who classify as middle class shrunk from 61% in 1971 to 50% in 2020[7].

This is running directly contrary to global middle-class growth, with the middle-class expected to grow to over 75% by 2050.

Lessons from history

International Monetary Fund data over the last 73 years shows that countries with greater levels of income and wealth inequality are inherently less stable. When the rich funnel their savings into property and financial assets, the economy over time becomes more volatile. The most robust and stable economies have a healthy middle class that is broadly consuming and injecting capital into the economy in a virtuous circle of growth. This is why economists talk about consumer confidence in the same way they talk about trade deficits. When wealth capture reduces the ability of spending to stimulate the economy, labour growth and wealth distribution, we need to look at consumption's role in economic stability.

People want to be productive, but they also need to feel valued and to believe that their hopes and aspirations have some validity. If large numbers of people are without jobs, without purpose, and without the prospect of personal fulfilment, but are living in a stratified society where they see the dreams and aspirations of only a few being realized, then those people naturally feel frustrated and angry.

"Research indicates that increased inequality can erode social cohesion, lead to political polarisation, and ultimately lower economic growth. The IMF's work on income inequality looks at how fiscal policy can help governments tackle high levels of income inequality and benefit economic growth."
—IMF Policy Paper (2018): IMF's Work on Income Inequality[8]

The key economic problem with rising inequality is that historically it drives long-term economic activity down and at the same time increases the likelihood of serious social division, further dampening economic confidence. Ray Dalio, billionaire founder of the world's biggest hedge fund, put it this way: "If I was the president of the United States, what I would do is recognize that this is a national emergency ... If you look at history, if you have a group of people who have very different economic conditions, and you have a long-term economic downturn, you will have conflict."[9]

This view is hardly revolutionary (excuse the pun). David Brin, a *New York Times* bestselling science fiction author, dreams and writes of societies that emerge 10,000 years from now, but that ability to project distant future social and economic models is based on historical precedents that he sees as being likely to inform future developments. Brin observes that if you look at Adam Smith's *The Wealth of Nations*, or even the Bible, you see that resentment between the rich and the poor is par for the course over the last few millennia. Conversely, there have also been periods where equality and robust middle-class growth have produced extraordinary economic results.

At the conclusion of World War II, wealth inequality in the United States was at an *historical all-time low*. The middle class had fully recovered from the Depression, and the austerity of the war had resulted in a strong savings culture. Good wages, full employment, and GI Bill-fostered competitiveness led to the most dominant economy that the world had ever seen. The civil rights movement was enabling the mobilization of poor black communities for the very first time. Housing construction boomed. Consumption skyrocketed, with nearly every home in America able to afford a refrigerator, washing machine, TV, or automobile. Everyone was an active participant in the US economy. Pop culture emphasized this in TV and radio. Capitalism was victorious! The American Dream was alive and it was the envy of every other nation on the planet.

Historians Ariel and Will Durant classified this type of social structure as a diamond-shaped socioeconomic model. In this model, a healthy and robust middle-class vastly outnumbered the poor and rich alike, and wealth distribution and wages followed a classic bell curve. The argument made by Brin and Durant, supported by centuries of historical data, is that these diamond-shaped systems produced not only the most optimistic and positive citizens, but also the most productive economies.

Today, the Chinese economy resembles this type of structure more so than the US economy. Economists today would argue that domestic middle-class growth is one of the keys to China's self-sufficiency as an economy, but also that the overall outlook is healthy because of significant investment being made in infrastructure and modernizing the economy through education, etc. Today, it's China that looks like a diamond-shaped model, not the US.

Examined historically, however, these diamond-shaped models appear unstable despite their tremendous success. Why? The Durant's argue that the so-called 0.1%, the feudal lords of history, aren't happy with more evenly balanced models of socioeconomics and seek to undermine or game them. They seek, through policy, lobbying and influence, a return to a pyramid-shaped model where wealth flows uphill and where the wealthiest citizens control the political discourse and economic trajectory of the nation.

"Adam Smith himself contended, in both **The Wealth of Nations** *and* **The Theory of Moral Sentiments,** *that a relatively flat social order—combined with lots of opportunities for the poor to get education, so the total number of competitors is maximized—can vastly increase the total number of people who get rich in the best way, by delivering innovative goods and services."*
—Contrary Brin Blog: David Brin,
"Class War" and the Lessons of History

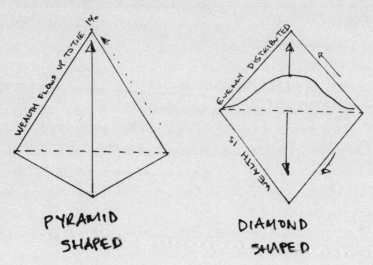

Figure 13: Pyramid versus diamond-shaped socioeconomic models.
(Adapted from Durant's *Lessons of History*)

Ariel and Will Durant demonstrated through historical examples that despite the assurance of conservative economists arguing for tax cuts on the wealthy and the empowerment of corporations and the 1% (who might "trickle down" their wealth), history has been shown to reward *exactly the opposite* in terms of pure economic growth. Strong middle-class social structures with broad national prosperity and economic growth have consistently outperformed those periods where inequality is rampant. Efforts to educate the poorest citizens pays off in spades, with greater competition, innovation, and broad consumption. Brin argues that the founding fathers in the US attempted to enshrine a set of levelling mechanisms at the constitutional level that would avoid the accumulation of wealth at the top of the pyramid. Societies where monarchies had ruled for centuries reacted by becoming more diamond shaped in modern times, but those economies still had policies that hurt the poor over time.

The simple reason that diamond shape structures are much better in terms of economic growth comes down to broader economic participation. Broad middle-class consumption is historically more effective at creating economic activity than the rich pooling assets and capturing more and more wealth in investments, property, and the stock market.

Logically, consumption correlates with income, so with real income remaining flat for so many for so long, it's easy to see why the US economy has also flattened. In the US, consumer spending contributes 69% of gross domestic product[10]. Hence, it is generally argued that tax breaks will stimulate an economy through an increase in personal spending. But when we look at the effect of tax cuts on different tax brackets we see why middle-class consumption growth is so fundamental to the economy.

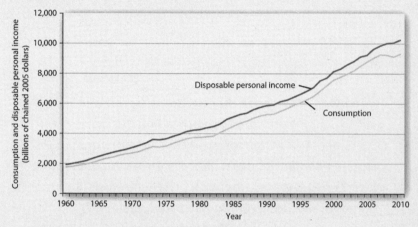

Figure 14: Consumption and disposable income.

Research by the Congressional Budget Office (CBO) and the Institute for Policy Studies found tax cuts varied considerably in their effectiveness, depending on which segment of society the cuts were aimed at. For the lower and middle-income segments, a reduction of tax revenue to the tune of $1 million accounted for seven new jobs being created. For tax cuts targeting the richest Americans, it was just four jobs per $1 million. Corporate tax cuts were even less effective. Between 2008 and 2015, publicly held corporations that paid less than 35% corporate tax reduced the number of jobs in the economy, while the overall economy grew employment by 7%. Instead, the most common side effects of lower corporate taxes were companies buying back their own stock and increased CEO pay[11].

Warren Buffet, one of the greatest modern American capitalists, argued that trickle-down economics had conclusively failed in an op-ed he penned for *Time Magazine* back in January of 2018:[12]

> *"Between the first computation in 1982 and today, the wealth of the [Forbes Richest] 400 increased 29-fold—from $93 billion to $2.7 trillion—while many millions of hardworking citizens remained stuck on an economic treadmill. During this period, the tsunami of wealth didn't trickle down. It surged upward."*
> —Warren Buffett, *Time Magazine*

Will Durant expressed his concerns about the potential effects of rampant inequality in his 1968 book *The Lessons of History*:

> *"In progressive societies the concentration [of wealth] may reach a point where the strength of number in the many poor rivals the strength of ability in the few rich; then the unstable equilibrium generates a critical situation, which history has diversely met by legislation redistributing wealth or by revolution distributing poverty."*
> —Will Durant, *The Lessons of History*

Durant argues that this is a well worn political cycle. The wealthy try to amass more and more wealth, destabilizing equitable wealth distribution, but when this creates the most acute forms of inequality, society bites back—either through legislative pressure or via revolution. Redistribution of wealth stabilizes society for a period of time, until the wealthy and powerful work out how to game the system back in their favour.

Looking back on the policies of Reagan, Thatcher and Clinton, we did indeed have legislated redistribution of wealth—it was just redistributed to the rich, not the poor. Following exactly the historical pattern that Durant recognized and one that is increasingly likely to lead to social revolution.

Inequality is an issue that has already delivered the populist movements and the greatest increase in protests by volume and number that the modern world has ever seen. It has worsened during coronavirus, making its potential for ongoing social and political destabilization even more acute. That is well before the worst impacts of automation and climate change have even had a chance to hit us.

Broad inequality must be addressed if we wish for any type of social coherence as these future crises hit. Addressing inequality proactively rather than waiting for it to bite is also the right thing to do.

Endnotes

1   Source: Public Religion Research Institute and *The Atlantic* "Beyond Economics: Fears of Cultural Displacement Pushed the White Working Class to Trump", May 2017.

2   See https://theconversation.com/who-exactly-is-trumps-base-why-white-working-class-voters-could-be-key-to-the-us-election-147267.

3   Credit: Midnight Oil, "Read About It" (1982).

4   Source: *Sydney Morning Herald* "Michael who? It's Gekko we're after" Phillip Coorey, 26 September 2008.

5   Source: *Business Insider*, "How billionaires saw their net worth increase by half a trillion dollars during the pandemic", 30 October 2020.

6   Source: Brookings Institute, Department of Homeland Security and Pew Research estimates (November 2019).

7   Source: Pew Research.

8   Source: International Monetary Fund: https://www.imf.org/external/np/fad/inequality/

9   Source: Bloomberg.com "Dalio Says Capitalism's Income Inequality Is National Emergency", 7 April 2019: https://www.bloomberg.com/news/articles/2019-04-07/dalio-says-capitalism-s-income-inequality-is-national-emergency.

10   Bureau of Economic Analysis. "National Income and Product Accounts Tables: Table 1.1.6. Real GDP," 2018 Dataset.

11   Source: Institute for Policy Studies (2018)—Seven Surprising Tax Facts for 2018.

12   Source: *Time Magazine*, "Warren Buffett Shares the Secrets to Wealth in America", 8 January 2018.

CHAPTER 2

HUMANITY'S TIPPING POINT

....

"Never let a good crisis go to waste."
—Sir Winston Churchill

....

Humanity has had its fair share of crises and disasters. Even though our historical record is limited, from Mount Vesuvius erupting and burying Pompeii in lava and volcanic ash in 79 CE through to the 2004 Boxing Day Tsunami that hit Southeast Asia, major global disasters, epidemics, and economic collapses are etched in our collective memory. True global disasters that have affected broad geographical regions and broad sets of the population remained fairly rare, however, at least up until the last century or so (not counting the Chicxulub comet, of course).

The Black Death was epoch defining, but it was made far worse by the advent of trading ships that traveled between medieval ports. With the emergence of air travel, higher levels of migration and technological advancements, the potential for pandemics to have broader geographical impact is magnitudes higher than it was before the modern age. With the rise of the global economy, we can add to that potential mix the nature of economic crises that can trigger global recessions with uncomfortable ease—as the saying goes, when China or the US market sneezes, the rest of the world catches a cold.

The nature of today's modern society means that pandemics, global financial crises and the effects of climate change are not only indiscriminate, but affect more of the globe than previous crises ever impacted ancient populations. It also means that global cooperation in dealing with such events is more critical than ever, at a time when deep ideological divisions make such cooperation seem impossible. Some might argue that globalism and things like modern technology are to blame, and that a more closed, independent society might fare better. However, no modern economy disconnected from the rest of the world is viable, it won't grow and thrive to benefit its citizens. Brexit may offer further proof for this over the coming decade.

Despite all of this history, as a species we are simply abysmal at planning our future or preparing for such crises. We do some disaster prevention planning at a civic level, but beyond budgets for emergency response capability, we tend to simply "hope it doesn't happen here". Economically we plan for growth, but in the Western world, we don't have a 20-year or 50-year infrastructure development plan, let alone a plan for what we'll do if cities like New York and Miami are inundated with rising seas. We've hardly changed the basic education curriculum since the beginning of the Industrial Age, and yet we're trusting it to train our children to live with robots and flying taxis. We kick the can down the road, leaving such problems to the next generation, when it is within our grasp today to build a better system for our children and grandchildren. Why?

The system today doesn't incentivize us to build a better future for our descendants. It incentivizes us to create short-term wealth. If there's no return on investment over the next couple of quarters, an idea struggles to get traction.

Are science and technology the answer?

Over the last 300–400 years as science and technology has advanced, we've seen a repeating pattern of debate—is this good or harmful for society? When it comes to the impact of Artificial Intelligence (AI) on our society, the same debate rages today. Will AI be beneficial for society? Which jobs will be impacted? Which economies will win out? Will it result in net positives or accelerate inequality?

AI has been a popular subject of science fiction for decades, but we are increasingly debating whether it will be good or bad, as the potential of AI looms ever closer. We have the likes of Elon Musk and the late Stephen Hawking discussing the almost apocalyptic potential of AI to destroy humanity. To be fair though, Musk has also said AI will eliminate the need to work and sees ways of us competing, such as Neuralink. Others, like Peter Diamandis and Ray Kurzweil, preach that AI will bring about a cultural and economic renaissance for mankind, leaving us with abundance, longevity and unimaginable collective wealth as a species.

We debate this potential future as if, somehow, we have a choice in respect to the adoption of AI.

Global capital markets are geared almost exclusively towards rewarding companies that differentiate through innovation and high levels of automation and productivity. It's largely an illusion that any nation can have isolationist policies that don't reflect the inevitable trend towards globalization, or that we can somehow ring-fence technology like AI, mobile commerce and social media in favour of bringing back the good 'ole days of yore when people worked automobile production lines by hand, and ploughed fields with horses. Globalization is largely a consequence of improvements in communications, producing increasingly connected markets and commerce. History shows us that in the last 250 years no industry or government has ever successfully slowed or stopped the impact of technology.

The most profitable companies in the world employ far smaller labour forces than the major blue-chip industrials did back in the day. As a result of continued productivity gains and the emergence of the GAFA/FAANG/BATX's of the world, tech companies can produce 10 times the profits that leading index companies of the 1960s did, and with far fewer employees. And yet machine learning is still in its infancy—imagine the impact when these technologies really take hold.

At the heart of the Fourth Industrial Revolution will be AI automating large swathes of society. Today the algorithms we are training are rapidly starting to compete with humans when it comes to any process that is repetitive in nature. Algorithms can now diagnose cancer and analyze x-ray and MRI results at levels competitive with human technicians. Complex financial algorithms have replaced tens of thousands of traders, financial advisors, credit risk officers and others. Soon robots will drive us in autonomous vehicles and deliver groceries, food and consumer goods, where previously humans were necessary. Researchers predict that in developed economies like the US this continued use of automation will impact at least half the workforce. In developing economies, it could be even higher.

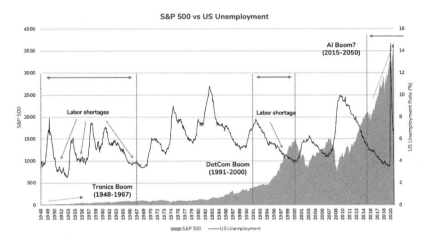

Figure 1: Labour shortages have previously accompanied technology booms.

Analysts are evenly divided on whether or not AI will result in the massive creation of new jobs or mass unemployment. For those that argue for the former, they cite the fact that in the "tronics" boom of the 1960s and the "dotcom" boom of the late 20th century, labour demand resulted in shortages of skilled workers to fill newly-created job profiles, and that the accompanying wealth created during these booms led to a range of service industries flourishing. McKinsey found that during the dotcom boom, as an example, the internet did eliminate some jobs where inefficiencies existed, but for every job displaced between 2.4 and 2.6 new jobs were created.

"A detailed analysis of France over the past 15 years shows that the Internet created 1.2 million jobs and destroyed 500,000 jobs, creating a net 700,000 jobs or 2.4 jobs for every one destroyed. This result is also reflected in [a McKinsey] survey of more than 4,800 SME's in the countries studied, which shows that 2.6 jobs were created for everyone that was destroyed... Further, companies that have fully integrated the technology and use it extensively create more than twice as many jobs as the average."
—"Internet Matters: The Net's sweeping impact on growth, jobs, and prosperity", McKinsey Global Institute, May 2011

What makes the coming AI boom fundamentally different from the tronics boom of the 1960s and the internet/dotcom boom of the late 1990s is that AI is a much broader realignment of core skill sets across the economy. Rather than being akin to the dotcom boom where the commerce layer was primarily affected, AI's impact on employment is much more like the emergence of the Industrial Age itself, where people were forced out of the farming and textile sectors into mechanization and industrialization. In the case of textile workers, in the space of 20 years most of those employed lost their jobs to steam-powered weaving looms, creating the Luddites of legend—the textile workers like Ned Ludd who attempted to sabotage the new automated looms because of employment losses.

While AI will definitely give birth to a whole raft of new jobs, it will displace core jobs in industries that have been relatively unscathed by technological development over the last 200 years, jobs where processes have remained largely human-based.

The World Economic Forum predicted in *The Future of Jobs Report* (2018) that this could go either way. If we prepare for this reskilling at an economy level supporting new training and programs that assist displaced workers, we could end up with a net gain: 75 million jobs displaced and 133 million new jobs emerging. But that is only if we take a very methodical approach to creating job programs, rethinking education and longer-term strategic policy development. If not, then in most economies, at least half of the pre-2020 jobs we knew will disappear by the 2040s.

The bigger change, which we will address in later chapters, is that this could mean a fundamental shift in the way we view work itself.

Assuming AI has even half the impact projected, there will be many jobs that are automated away, those who are gainfully employed will probably be working less, and it is fairly obvious that AI will not lead to wage growth or improve the lot of the poor and middle-class in the medium-term. The economic uncertainty we see behind the populist movements and protests around the world could conceivably be much worse.

Then throw in climate change. When researching projections on the potential impact of global warming, we looked at the scientific research but we also made sure to include work done by non-scientific bodies like the US military, the United Nations, the largest reinsurance companies in the world, studies from the world's largest energy companies, projections from hedge funds and financial institutions, urban planning departments on likely affected cities, government departments, private think tanks and research bodies. If you still don't believe that anthropogenic climate change is real, you are in a minority today. According to a 2019 poll taken by the Recycling Partnership, 96% of Americans are worried about climate change in some form, with three out of four now believing that it will eventually result in the extinction of humanity. Regardless, data based on the most primitive of measures (using a 17th century thermometer and measuring coastal water lines) show us that the planet is warming, and seas are rising—these basic metrics are irrefutable. The only real debate appears to be about how fast it is happening and whether mankind is to blame.

When looking at social cohesion on a global scale, the potential for climate change to wreak havoc, accentuate uncertainty and to introduce risk is unparalleled (see the box on *Climate Change Chaos* at the end of this chapter).

Permission versus precaution

When we look back at the history of technology and the disruptions it has wrought over the last few hundred years, we learn that while some have succeeded for a short-time in suspending the progress of various technologies, that is all that they have ever been able to do. At no time throughout history have humans successfully halted the march of a single technology and its impact on society.[1] The same could be said of climate change.

Rather than debate the pros and cons of AI and climate change then, history tells us that our time would be far better spent in preparing society for their inevitable impact and doing all we can to smooth that transition. How do we properly prepare society for the most disruptive technology we've ever seen? How do we modify our behaviour to ensure that future pandemics don't have the economic fallout we've seen during COVID-19? How do we produce global policies to mitigate the impact of displacement of coastal populations from rising seas and the flood of immigration that will follow? What impact will these issues have broadly on society, governance, economics and politics?

The introduction of the steam machine and electricity were game changers for the world, kicking off the Industrial Revolution. Transcontinental railroads, telegraphs and interstate highways allowed businesses to thrive by creating national value chains in the United States in the 1860s. The discovery of oil and gas, along with the creation of the combustion engine, were responsible for massive wealth creation and economic growth in the early 20th century around what we today call the petro-dollar.

Many countries, especially in the developing world, are in the midst of major technological transitions today[2], bringing about transformations in the way a society operates, functions, and performs both technologically and socially. These developing economies are now faced with absorbing technological change at even faster rates than in the developed world during the 20th century.

The key question is this: if large groups of individuals or society as a whole decided to reject certain technologies or ideologies (such as climate change), does that have a negative impact on society? If the US had rejected the steam machine, railways, telegraphs and the combustion engine, would it be the world's largest economy today? Would society be better off?

In his book *Permissionless Innovation: The Continuing Case for Comprehensive Technological Freedom*, Adam Thierer posits that humanity is on the brink of the biggest advancements in recorded history. He argues that this leap in technology is only really possible if society accepts a largely "hands-off" approach to the development of new technologies. Thierer argues that innovation should be largely permitted by default.

The flip side to this ideology is the "precautionary principle", which maintains that new innovations should be curtailed or disallowed until their developers can prove that they will not cause any harm to society. Thierer argues that the precautionary approach is harmful to society as it dramatically reduces economic competitiveness, output and market gains. If we restrict the deployment of AI, as an example, until we can prove that it won't take jobs from humans, we're increasingly unlikely to implement that technology. However, if we permit AI to be unleashed on society without any protections or precautions, we're also likely to experience unintended consequences and see massive disruption. At a minimum, there will be a large-scale employment impact. How do we manage these two opposing risks?

Our view of risk can also be biased by overzealous and sensationalist reporting or based on the social media bubble that we live in. For example, when it comes to the coronavirus pandemic, in December of 2020 the Journal of American Medical Associations reported that COVID-19 had become one of the leading causes of death in the United States, above heart disease and cancer. Media reporting and social media conjecture on coronavirus and terrorism, for example, far outweigh the actual risks to society by order of magnitudes.

"The news media dutifully report each day's increase in new cases and deaths but putting these numbers in perspective may be difficult. The daily US mortality rate for COVID-19 deaths is equivalent to the September 11, 2001, attacks, which claimed 2988 lives, 1 occurring every

1.5 days, or 15 Airbus 320 jetliners, 2 each carrying 150
passengers, crashing every day."
— "COVID-19 as the Leading Cause of Death in the United
States", JAMA, 17 December 2020 (Woolf et al)

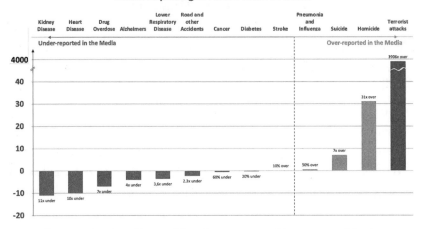

Figure 2: Based on Shen et al (2018) comparison of 2018 causes of death,
versus share of media coverage.

How can we filter our own "mainstream media" and social media circles to accurately assess personal risk to ourselves, our family and our friends? If we could solve this, we'd either be very rich or vilified by every special interest group that prefers to live in their own information bubble. It is now understood that social media played a critical role in advancing conspiracy theories around Trump's 2020 Election loss, which subsequently led to an attack on the US Capitol Building on 6th January 2021. This is clearly a bad outcome for society. Should we ban social media as a result? Most would argue no.

The rise of Trump and other populists, the rise of anti-vaccination conspiracy theories and theories on the origin of coronavirus are clearly tied to social media platforms and broad tribalism. What are the costs of these feedback loops on society? Ironically, the spread of misinformation via mainstream and social media are both arguably causes of significant economic damage over time.

Data transparency should be required in a society driven by data. Trusted sources and consensus are useful mechanisms for creating broad acceptance. Ultimately, we must have a better way of filtering out misinformation versus verifiable facts and data. There's no room for alternative facts in a coherent unified society. False data must be excised. Science has provided mechanisms for this built around peer review of data by qualified peers; but in the case of mainstream or social media this is obviously not possible. AI, however, may give us the means to tag information by its associated data and factual resonance in the future.

Let us look at two precedents where the rejection of facts has resulted in a significantly negative economic impact.

Anti-vaxxer movement: US$20bn+ year and one million deaths per year

The rejection of vaccinations in economies like the US has brought back diseases considered to be largely eradicated, such as measles. The US' Centers for Disease Control and Prevention (CDC) found that measles outbreaks in 2019 (defined as three or more cases) hit their second-highest level since the country effectively eliminated the disease in 2000. The highest reported number of measles cases since elimination was 667 in 2014. This correlates with the anti-vaccination movement that arose because of a Lancet report in 1998 that has since been widely discredited as a result of fabricated data. Measles cases globally have jumped 30% since 2016, according to the WHO, and measles was listed as one of the top 10 health threats in their 2019 annual report. As the world rolls out vaccines to halt the spread of COVID-19, we see large groups of individuals refusing to participate in these broader health initiatives due to widespread disinformation campaigns.

Let's look at the historical performance of vaccines through their track record on an economic and performance basis:

1. **Smallpox**

   The first recorded instance of smallpox was in 1350 BCE. Smallpox accounted for total cumulative deaths of 300–500 million before the vaccine was made available. In 1967, 10–15 million cases per year were recorded with a 17% fatality rate. Of those that didn't die from smallpox, 100,000 per year were left blind. From 1958 to 1979, a global vaccination campaign was initiated by the WHO. There have been no smallpox deaths in the last 25 years.

2. **Measles**

   Measles emerged somewhere in the 11–12th centuries. In 1657 the first recorded case of measles was recorded in Boston, USA. More than 200,000 cases of measles were recorded during the American Civil War, resulting in more than 500 deaths. Prior to the emergence of a vaccine in the US, 95% of the population was infected by age 15. Throughout the 20th century annual averages recorded 500,000 cases, with 500 US deaths per year (historical average 2.83 deaths per 1,000 people). From 1978 to 2000, a global vaccination campaign was embarked upon. By 2000–2017, measles was virtually eliminated in the US (2–3,000 mild cases per year in non-vaccinated households), and globally measles cases fell 93% between 1990 and 2016. In 1998, *The Lancet* published a paper by Dr Andrew Wakefield linking the MMR vaccine to autism. In 2010, The Lancet retracted the paper after evidence emerged that the research was fabricated. By 2019 previous progress had been reversed, with 140,000 cases worldwide (the worst year on record for measles since 1950).

3. **Polio**

   Hieroglyphic reliefs dating back to 1400 BCE in Egypt show polio symptoms. The first polio epidemic in the US was recorded in 1875. In 1916, a polio epidemic in New York resulted in 6,000 dead and 20–30,000 individuals permanently paralyzed. It resulted in the shutdown of pools, amusement parks, public libraries and public parks due to fear of contagion. Iron lung treatment was introduced in 1928, at the cost of $1,500 per unit (about the cost of an average family home at that time, but around $23,000 in today's terms). The use of the iron lung represented the most effective treatment for polio until a vaccine was made available in 1955. Prior to the vaccine, the world recorded 350,000 cases annually up until 1988. Today that is less than 500 a year. Dr. Jonas Salk, the vaccine inventor, is still considered by most to be a national hero in the United States. Thanks to the recent work of the Bill & Melinda Gates Foundation, only three countries remain that have not eliminated this disease completely.

4. **Typhoid**

   Two US presidents, namely William Henry Harrison and Zachary Taylor, died of typhoid. During the Civil War, 80,000 Union soldiers died of typhoid, more than the number that died in battle. During the late 1800s the entire U.S. Army was vaccinated, eliminating typhoid as a significant cause of death during World War I. Today it is estimated that 11–20 million people annually contract typhoid due to poor-quality water sources, and this still results in 120,000–160,000 annual fatalities. While vaccines exist for typhoid, Salmonella Typhi, the bacteria that causes typhoid, can be passed on by wastewater mixing with other water sources, for example.

Vaccines have a history of incredible life-saving successes resulting in the complete reversal of major epidemics for dozens of diseases. Between 2010 and 2015 alone, vaccines prevented at least 10 million deaths[3]. Taking a historical view, cumulatively it's likely that more than a billion people have been saved from either death or long-term debilitating conditions as a result of vaccines already in use. Arguably this is many trillions of dollars' worth of benefit to economies overall as the prevention of mass deaths, elimination of massive medical and treatment costs, reduction of economic losses, etc, have all resulted as flow-on effects.

However, the recent rise of the anti-vaxxer movement is likely to reverse much of this progress. The cost to the global economy of the anti-vaxxer movement is estimated to be as high as US$2–5 billion, and upwards of one million avoidable deaths annually. Surely, educating people on the science and the incredible success of vaccines of the past is one way to combat this misinformation.

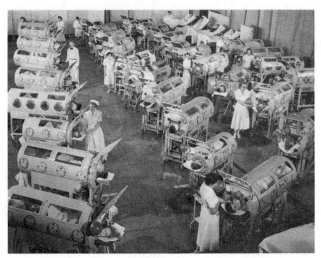

Figure 3: The use of iron lungs prior to the polio vaccine.

Vaccine-preventable diseases among adults cost the US almost $9 billion per year in treatment. Vaccination prevents 2–3 million deaths a year today, but it could prevent a further 1.5 million a year if coverage improved and we reversed the trend of vaccine hesitancy that has emerged over the last two decades. Measles cases in Europe are at a 20-year high and 72 children and vulnerable adults died of it last year, where deaths had previously been eradicated for over a decade. New UK National Institute of Health Research shows that the total cost of the 2012–13 measles outbreak in Merseyside (£4.4 million) was more than 20 times the cost of the vaccinations that could have prevented that same outbreak (£182,909). These costs incorporated lost employment from having measles or looking after someone with it, and accounted for approximately 44% (£2 million) of the total.

In April 2020, a Kaiser Family Foundation study projected that the cost of treating just COVID-19 cases for the uninsured would range from $13.9 billion to $41.8 billion ("Vaccine Refusal Will Come at a Cost—For All of Us", *The Atlantic,* 10 April 2021).

Oh, and that controversial MMR Lancet report from 1998 that led to rejection of the measles vaccine? Recent research actually suggests gut microbiome is responsible for up to half of Autism Spectrum Disorders[4]. Despite this, millions still today believe that vaccines are harmful.

The long-term costs of dismissing the science around vaccines is likely to be in the trillions of dollars globally. What could that injection of capital do for other areas of the economy? The rejection of the COVID-19 vaccine itself will do further unnecessary damage to economies already affected negatively by the pandemic. We clearly need to do a better job at educating people on how vaccines work, and giving them unbiased access to facts that are understandable and defensible.

Climate change inaction: 10–60% of global GDP by 2050

When looking at impact on the global economy, there are few potential issues that can match the scale, persistence and systemic risk associated with climate change.

The US, Great Britain, and Australia lead the world in climate change deniers. In contrast, if you live in China, Argentina, Italy, Spain, Turkey, France or India, you are likely one of the more than 80% of the population who accept climate change as a fact. According to Morgan Stanley, climate disasters have cost North America $415 billion alone in the last three years, much of that due to wildfires and hurricanes which are likely linked to climate change. In 2017, Texas's estimated losses from Hurricane Harvey were $125 billion. Hurricane Sandy caused about $71 billion of damages in 2012.

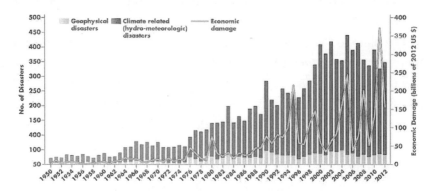

Figure 4: The Rising Cost of Global Warming: Increasing Frequency and Cost of Natural Disaster. (Source: EMDAT (2017); OFDA/CRED International Disaster Database)

The economic impact of more severe and intense weather is already having a marked impact on the global economy. In Australia, the Climate Council of Australia estimates that the loss of the Great Barrier Reef to bleaching from higher ocean temperatures will result in losses of $1 billion in tourism activity annually by 2025, and more than 10,000 jobs.

On 16th January 2019, the last four Federal Reserve heads, 15 former leaders of the White House Council of Economic Advisers, and 27 Nobel laureates (mostly in economics) signed a letter endorsing a gradually rising carbon tax whose proceeds would be distributed to consumers as "carbon dividends" or offset.

In calculating the costs of inaction against a changing climate, the economics are a little tricky. There are the obvious costs of changing weather and damage to infrastructure and the economy, and catastrophic events like food shortages or climate displacement leading to mass refugee crises. But by upgrading energy infrastructure, for example, while the costs initially are high, the long-term efficiencies in energy savings, lower cost, and greater reliability will have a positive economic impact. Climate shift will also see greater inequality as the wealthy would likely be able to bear the costs of relocation and adaptation more easily.

Over the next 20–50 years, the impact of climate change will get more severe, and the economic impacts are likely to be the most expensive in history. In the final year of World War II, about 40% of the world's

GDP was spent on the war effort. For climate change that could be as much as 60% of global GDP by 2050 alone. Reluctance to deal with climate change today doesn't mitigate the need for massive spending at some point. Ironically, when we finally have to deal with climate change it might produce massive economic impetus for those that survive the initial disasters, failed economies and inundation by both rising seas and floods of immigration. For the second half of the century, as we deal globally with climate change, investment in adapting and future-proofing humanity and civilization will likely lead to an economic (engineering and infrastructure) boom even greater than the post World War II years, alongside massive social disruption. But the human toll will likely be well north of a billion people. That's a hell of a trade-off for inaction. The economic effects of climate change could be far worse than we anticipate. They could include:

- Loss of almost 600 major coastal cities due to sea rise by 2050
- Rice, wheat, potato and maize production could retract 12–25% by 2050
- 360 million to 1 billion climate eco-refugees could be created by sea-rise levels and agricultural failures by mid century
- The complete collapse of the $6 trillion-a-year global insurance market as rolling climate disasters make insurance coverage untenable
- Heat related deaths will increase 250–300% by 2050

In the meantime, the historical resistance and broad inaction of governments globally will mean negative economic and social impact such as the world has never seen. That's already beyond question. If looked at in the aggregate, it appears to be absolute insanity that humanity is not completely repurposing the entire global economy to fix this as we write. It doesn't make sense that this isn't the number one news story on every newspaper front page and cable news TV show every day of the year. Why isn't it?

Globally, fossil fuel subsidies remained massive counterpoints to climate progress at $5.2 trillion (6.5% of GDP) in 2017. The largest subsidizers were China ($1.4 trillion), United States ($649 billion), Russia

($551 billion), European Union ($289 billion) and India ($209 billion). Ironically, unsubsidized fossil fuel pricing in 2015 would have lowered global carbon emissions by 28% and fossil fuel air pollution deaths by 46%, while increasing government revenue by 3.8% of GDP. When it comes to electricity generation, fossil fuel subsidies have remained more than double that of renewable energy the last few years.

Estimates for global fossil-fuel consumption subsidies

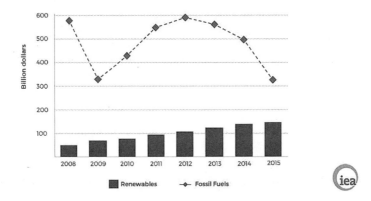

Figure 5: Global fossil fuel vs renewables subsidies (Electricity generation only).
Source: IEA

From 2000–2016 in the US alone, the fossil fuel industry spent nearly $2 billion lobbying to prevent climate action.[5] This doesn't include the money spent on corporate reports and science refuting climate change or global warming over the decades. It doesn't include political donations to politicians for their continued support of climate inaction. It doesn't include billions in fossil fuel subsidies.

Is there a better argument to demonstrate that capitalism and free markets can be manipulated into serving an industry over humanity? We think not.

Why so much technology trepidation?

The resistance to advances in technology and science has arguably threatened the world with extinction, shortened the life spans of millions (because of disease, pollution, poor sanitation, etc.), limited economic potential and has held back entire classes of people with limited access to basic services like healthcare, banking, and education. At the same time, technology has increased the potential for inequality, created vast economic divides, increased political and social tensions, fueled racism and hate crimes, and been responsible for many of the problems we face today.

Gene therapy will soon give us the tools to eliminate genetic disorders and inherited diseases from our genome, but we could see the rise of designer babies with the ethical conundrums that accompany that. We will soon see the first humans on Mars, but many will argue that spending billions on space when we've got so many problems at home is a travesty. Today, 72% of Americans can order from Amazon and have their order delivered the very next day; but there are still over a billion people on the planet without access to clean water or basic sanitation. We have smart watches that help predict if we're going to have a heart attack, but this will involve massive breaches in respect to personal medical data and insurers who refuse to give you insurance based on such sensor data. We'll have autonomous vehicles that eliminate traffic accidents and fatalities, but we'll put millions of truck drivers out of work.

It seems like there's no easy answer. If we do adopt technology, we'll certainly face challenges and there will be unrest and displacement; but if we resist these advances, we will ultimately do even more damage, especially to the future that our children and grandchildren will inherit.

When we focus on what could go wrong with technologies like, for example, AI or gene therapy, the discussion often revolves around this conundrum: should we adopt this technology, or should we prevent it from doing possible harm?

The problem with this argument itself is that history teaches us we will always, inevitably, take up the technology. The rejection of the 8-track and Betamax aren't examples of society's broader rejection of technology. Economies that reject technology at scale always find themselves trailing in terms of competitiveness and ultimately have to adopt the technology that they shunned so as to remain competitive. Skill sets change, industries wane. We can now see a time when the use of coal and oil will be eliminated across the planet, despite the long-term efforts of politicians, industry lobby groups and unions who have fought changes in this space for decades. The movie and recording industries resisted the likes of Napster and BitTorrent, spending hundreds of millions attempting to prevent the download of media content, and they failed. Today streaming services are major media players.

Seen in a historical context, society would be much better off attempting to negotiate an orderly transition around such technologies, and not resisting them. We could be creating job training programs to transition coal miners into other industries, even to the mining of other resources like crystalline silicon, or retooling the energy industry towards renewables and energy storage systems. Especially when solar energy is cheaper than coal or gas generation in every major economy in the world[6]. It's inevitable that over the next 20 years the entire planet will be biased towards these renewable energy systems, so why not incentivize this?

What we end up with is various potential futures. At one end of the spectrum you have maximum resistance to change, and at the other an optimal path to adoption of new breakthrough technologies and adaptation to a changing world. At the same time, we could have clear policy direction, efforts at mitigating climate change, improving equality and inclusion and reinforcing strong ethical standards; or we could have political obfuscation, ignorance of basic scientific data and a total hands-off approach.

The task before us is anticipating what sort of society will evolve out of this melting pot of possibilities. We are going to attempt to answer that, and in doing so map out the most likely path or range of possibilities that the world is stumbling into. When we talk about a world of technosocialism, we're not talking about political strategies to change political philosophies; we mean a combination of pressures leading to a global movement that will reshape the societies we live in and the way we live.

The most urgent of these issues is the impact of climate change. If humanity as a species is to survive climate change, we will have to create entirely new industries dedicated to undoing the large-scale damage of carbon emissions, extracting pollutants out of our air and water, stopping the burning of fossil fuels and about deforestation of the planet. We must absolutely do this for future generations, but instead we are arguing what level of carbon output is politically tenable. A hundred years from now no human living on the planet will be arguing about coal miners losing their jobs, or car manufacturers not being able to emit pollution from internal combustion engines. Today we debate this while we accelerate the worst-case scenario for our grandchildren.

If humans had evolved to view resource allocation differently, if money and currency didn't exist, it's plausible in some alternate universe that climate change would never be a threat. But it is, so we have to fix the system that produced it. Is that too idealistic, too naïve? Giving our grandchildren a future where their lives are not defined by our poor policies and planning is the only responsible path in an ethical and inclusive society.

It's crunch time. You are either part of creating the solution, or you are endorsing the problem. You are either for the future of humanity as a species or you are married to political and economic ideologies created in a vastly different time that will kill more people than all the wars in history.

This is humanity's tipping point. Where do you stand?

Endnotes

1   We're not talking specific technologies over another (i.e. Betamax vs VHS), rather the trend for adoption of major technologies themselves—e.g. the telephone, television or the internet.

2   Rock, M.T., & Angel, D. (2005). *Industrial Transformation in the Developing World.* Oxford, UK: Oxford University Press.

3   Source: WHO: https://www.who.int/publications/10-year-review/vaccines/en/.

4   Source: Nature Scientific Reports, "Long-term benefit of Microbiota Transfer Therapy on autism symptoms and gut microbiota", Volume 9, Article number: 5821 (2019).

5   Source: "The climate lobby: a sectoral analysis of lobbying spending on climate change in the USA, 2000 to 2016"; Drexel University study Brul et al, 19 July 2018.

6   Source: International Energy Agency report "World Energy Outlook 2020" (October 2020).

CLIMATE CHANGE CHAOS

As climate change accelerates without being addressed, the following potential outcomes become increasingly likely:

Effects of sea level rise

Between 1900 and 2016, sea levels rose by 16–21cm (6.3-8.3in)[1]. Satellite radar measurements taken between 1993–2017 revealed a rise of 7.3cm (3in) over that period. One of the reasons estimates have increased over the last 20 years is computing power, satellite imaging and data collection have improved dramatically, and hence models to predict future sea level rises have become more accurate. The Intergovernmental Panel on Climate Change (IPCC) predicted a high-end estimate of 60cm (2ft) back in 2007, but they revised that in 2014 to 90cm (3ft). The more extreme estimates based on Antarctic Ice melt (with ice shelf collapse displacing into the ocean) are in the range of 200–270cm (6.6–9ft). The most conservative estimates still put sea level rise at 2.3 metres (7.5ft) over the next two centuries.

Most of the rise in sea levels have and will occur through thermal expansion of the ocean (42%) and glacial melt (21%). Both of these are clearly accelerating and are easily measurable.[2] The likelihood that estimates will be revised downward in the future seems low. Arguing over whether man caused this warming is largely irrelevant—it's happening, and we need to mitigate the risks involved.

The outcomes of sea rise are varied. The most obvious are widespread coastal flooding, higher storm surges and more dangerous tsunamis, displacement of populations, loss and degradation of agricultural land, and more regular damage in the multi-billion-dollar range to cities, such as what was seen in New York during superstorm Sandy, or New Orleans after Hurricane Katrina. The less obvious effects are ocean temperature and acidity radically affecting aquatic ecosystems, and a subsequent decline of somewhere between 6–11% of total global fish catch. The most concerning impact of sea rise is displacement of large populations of coastal dwellers.

The unintended consequences of sea rise that are already starting to appear are still extremely concerning. Of the top cities in the world affected by sea rise, many are in the three largest economies of the 2050s—China, USA and India. Miami is the most economically exposed city in the world in respect to sea rise and has a population of close to half a million people today, and while Governor Rick Scott sought to ban the terms "climate change", "global warming" and "sustainability"[3], property developers and Miami's mayor weren't as skeptical.

Table 1: Cities projected to be most affected by future sea rise. (Nicholls et al 2007, OECD)

Rank	Country	Urban agglomeration	Exposed assets—2007 ($billion)	Exposed assets—2070 ($billion)
1	USA	Miami	416.29	3,513.04
2	China	Guangzhou	84.17	3,357.72
3	USA	New York-Newark	320.20	2,147.35
4	India	Calcutta (Kolkata)	31.99	1,961.44
5	China	Shanghai	72.86	1,771.17
6	India	Mumbai	46.20	1,598.05
7	China	Tianjin	29.62	1,231.48
8	Japan	Tokyo	174.29	1,207.07
9	China	Hong Kong (SAR)	35.94	1,163.89
10	Thailand	Bangkok	38.72	1,117.54
11	China	Ningbo	9.26	1,073.93
12	USA	New Orleans	233.69	1,013.45
13	Japan	Osaka-Kobe	215.62	968.96
14	Netherlands	Amsterdam	128.33	843.70
15	Netherlands	Rotterdam	114.89	825.68
16	Vietnam	Ho Chi Minh City	26.86	652.82
17	Japan	Nagoya	109.22	623.42
18	China	Qingdao	2.72	601.59
19	USA	Virginia Beach	84.64	581.69
20	Egypt	Alexandria	28.46	563.28

Starting a few years ago, property developers rapidly started buying up properties in one of the poorest neighbourhoods in Miami—Little Haiti. Traditionally home to working class and immigrant families, Little Haiti also has the highest elevation in Miami City, ranging from 2.13–4.2 metres (7–14ft) ASL, compared with 0.9–1.2 metres (3–4ft) in places like Ocean Drive today. With sea rise of one metre (3ft) expected in the next 20 years, this clearly represents a problem to property developers, so they started buying up large swathes of property in Little Haiti[4]. This has created gentrification of one of the last affordable neighbourhoods in Miami, forcing working class families to move out of the city seeking affordable rents, but increasing their cost of living. Think about all the wealthy retirees who have relocated to ocean-front properties over the last few years, and who have had to face a 350% increase in flood events and saw 65 hours of flood levels above the 30cm (1ft) mark in just the last 12 months[5]. Governor Scott might be in denial, but that isn't a workable plan for Miami residents as sea rise continues.

Mass migration

Approximately 1 billion people occupy land that lies less than 10 metres (32ft) above current high tides, 250 million below a one-metre rise in sea levels. More accurate modeling based on NASA's SRTM (Shuttle Radar Topography Mission) digital elevation model utilizing neural networks has yielded very concerning data. A study published in *Nature* in 2019 concluded the following:

"Sea levels projected by 2050 are high enough to threaten land currently home to a total of 150 (140–170) million people to a future permanently below the high tide line, or a marginal increase of 40 (30–60) million. Total and marginal exposure each rise by another 50 (20–90) million people by end of century. A total of 360 (310–420) million people are on land threatened by annual flood events in 2100, or an extra 110 (60–170) million beyond the contemporary baseline."

—"New elevation data triple estimates of global vulnerability to sea-level rise and coastal flooding", by Kulp & Strause, *Nature*, October 2019

The study concluded that 300 million people could be forced to relocate as a result of sea rise and coastal flooding as early as 2050, rising to 480 million by 2100. This is based on us hitting Paris accord emission levels, but with Antarctic instability. Best case, 150 million.

Now think about the fact that the Syrian Civil War displaced upwards of 13 million people, 6.5 million people of whom are thought to have fled the country. In 2015 this led to a peak of 1.3 million first-time migrants entering the EU. Now imagine that on a global scale multiplied 10- or 100-fold. Shutting borders down or wiping our collective hands of the problem is simply no longer viable. What's our global plan for absorbing 150–300 million displaced climate migrants? We don't have one.

Agriculture yield

Critics of anthropogenic climate change (man-made climate shift) science will often argue the climate has always been changing. While that is true, it is the extraordinary rate of change that makes projected impact assessments most concerning. While mankind has been experimenting with agriculture over tens of thousands of years, plants have been bred to maximize their suitability to specific regions and areas. For each plant variety, there are optimal temperatures for growth and crop yields. As temperatures rise or drop, growth will vary. Typically, this range of optimal temperatures is quite narrow, meaning that crops will fail at temperatures outside this zone unless moved to areas that fit within the typical profile.

Corn, for example, will start to fail to produce at temperatures above 35°C (95°F); soybeans above 38.8°C (102°F). A 2008 study published in *Science* suggested that 30% of maize (corn) could be lost across southern Africa by 2030. For South Asia, rice, millet and maize could similarly be affected. Roughly 83% of consumable food calories come from 10 global crop types: maize (corn), rice, wheat, soybeans, palm oil, sugarcane, barley, rapeseed (canola), cassava and sorghum.

The biggest problem humanity faces in respect to food production is that population increases demand greater food production, at a time when climate change is set to significantly reduce production yields. Decreases in consumable food calories are already reported to be occurring in roughly half of the world's food insecure countries. For example, in India, food calories have declined by 0.8% annually and in Nepal they have fallen by 2.2% annually. In 2014, for the first time in more than 30 years, the number of undernourished people in the world increased. By 2017, the Food and Agriculture Organization estimated 821 million people are now undernourished, an increase of 40 million in just three years. Incidentally, according to the USDA, 37 million Americans struggle with hunger daily.

We can mitigate some of this risk by moving crops to areas that were once too cold to support those same plants. However, bear in mind that today many of these areas are simply not set up to support the sort of increased demand in crop yields that will be required. And areas subject to temperature increases will see many farmers and crops displaced. While wine is produced across the globe, France, Italy and Spain account for 50% of global production and have a disproportionately large impact on global supply. In 2017, the Bordeaux region suffered a 90% crop loss, the worst crop since 1945.

Endnotes

1   USGCRP (2017). "Climate Science Special Report. Chapter 12: Sea Level Rise". science. globalchange.gov.

2   Ocean temperature buoy networks and basic glacial runoff assessments are very basic science.

3   Source: Miami Herald, "In Florida, officials ban the term 'climate change'", Tristram Korten, 8 March 2015: https://www.miamiherald.com/news/state/florida/article12983720.html.

4   CNN, "Miami's Little Haiti wasn't a target for developers. Until the seas started to rise.", Bill Weir, 12 July 2019: https://www.cnn.com/2019/07/11/us/miami-little-haiti-climate-gentrification-weir-wxc/index.html.

5   Washington Post, "Sea level rise is combining with other factors to regularly flood Miami", Matthew Cappucci, 8 August 2019: https://www.washingtonpost.com/weather/2019/08/08/analysis-sea-level-rise-is-combining-with-other-factors-regularly-flood-miami/.

CHAPTER 3

THE TECH TRILLIONAIRES

"A world where 1% of humanity controls as much wealth as the bottom 99% will never be stable."
—President Barack Obama in his departing speech to the UN General Assembly, September 2016

If we imagine wealth as a funnel, it's easy to see why billionaires and future trillionaires have wealth that appears to grow almost infinitely. Some billionaires whose fortunes are tied to a single business, for example, do see their wealth decline in the event of an industry downturn or problems that are specific to that business, but for most their portfolios continue to accumulate more and more wealth every day. For many billionaires they simply could not spend or give away their wealth fast enough to end up with nothing—or even half of what they have today.

In 2000 Bill and Melinda Gates established their foundation and gave an endowment of 46.8 billion US dollars. At the time he was the richest person in the world, being worth $60 billion, and the endowment represented almost 80% of his wealth. Today, even after that endowment, Bill Gates is worth $135 billion. In 2020 MacKenzie Scott, co-founder of Amazon and ex-wife of Jeff Bezos, gave away $6 billion to charities and philanthropic organizations. During 2020 alone, her net worth increased by $23.3 billion.

As billionaires deploy their wealth in investments, new businesses, venture capital and private equity funds, they end up widening their ability to capture new wealth. This in turn makes them richer and richer. This wealth capture means today that the world's largest financial resources are unevenly allocated.

Case in point: let's look at the impact of the coronavirus pandemic on the richest and poorest of our fellow humans.

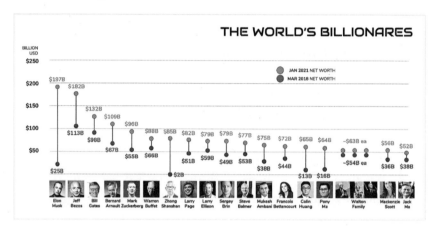

Figure 1: Gains in net worth for the world's richest billionaires.
(Source: Forbes, Seeking Alpha, Annual Reports, etc)

The world's top 25 richest individuals accumulated US$800 billion of wealth between March 2018 and January 2021[1]. Of the 2,189 billionaires across the globe, PWC calculated that collectively they grew their wealth by $1.9 trillion over the course of 2020[2]. That means that for the first time in history the world's billionaires collectively hold more than $10 trillion overall in wealth, up from $8.9 trillion at the end of 2017.

On the flip side, the World Bank warned that COVID-19 would result in 150 million persons slipping into "extreme" poverty by the close of 2021[3], up from around 80 million in 2019. This is the first increase in extreme poverty levels that we've seen globally in more than 20 years, a sobering statistic.

The World Social Report published by the United Nations Department of Economic and Social Affairs (UNDESA) at the start of 2020 showed that income inequality had increased in most developed countries since the 2008 financial crisis. For more than 70% of the population, inequality was already impacting their day-to-day quality of life before the pandemic hit. As noted earlier, the impact of both climate change and the coronavirus pandemic has already affected lower and middle-income households much more significantly than those at the top of the wealth spectrum.

A survey of 37 countries conducted by the Save The Children Fund found that three-in-four households suffered declining income since the start of the pandemic. We have observed at least four ways the COVID-19 pandemic has increased inequality across the globe:

1. While higher-paid workers work from home, lower-paid blue-collar workers typically do not have the same flexibility; thus they have to absorb higher costs just to keep working, even when their real wages have likely suffered.
2. A higher share of low-to-middle income (LMI) workers are employed in essential services like nursing, policing, teaching, cleaning, sanitation and retail stores, where they experience a higher probability of contacting an infected person.
3. LMI workers are significantly more represented in sectors where lockdowns or travel restrictions have had a negative impact, such as the hotel, restaurant and tourism industries.
4. The pandemic has accelerated increases in extreme poverty and inequality when comparing developed and developing nations, especially where we've seen those developed nations create broad economic stimulus programs that have assisted businesses and provided social safety nets for unemployed citizens.

As the long tail of the pandemic plays out, it's clear that inequality is going to continue to be a contentious issue. The medium-term issue is that the markets appear to be going through some pretty critical structural changes around technology that could make resolving inequality even more difficult in the future.

Technology reframes markets and wealth

Nine out of the ten most valuable companies in the world today are technology companies. In order of market capitalization as of January 2021, these were:

1. Apple $2.55 trillion
2. Saudi Aramco – $1.75 trillion
3. Microsoft – $1.7 trillion
4. Amazon – $1.6 trillion
5. Delta Electronics (Thailand) – $1.4 trillion
6. Alphabet (Google) – $1.2 trillion
7. Tesla – $834 billion
8. Facebook – $757 billion
9. Tencent – $738 billion
10. Alibaba – $620 billion

The only company in the top 10 most valuable organizations that isn't a tech company? Saudi Aramco. It's no surprise, then, that half of the richest 25 billionaires on the planet are also involved in technology-led companies.

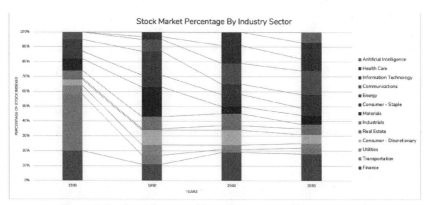

Figure 2: Stock market capital by industry sector. (Source: various)

In less than a decade the five largest tech stocks in the US have increased their value by $6.4 trillion, while during the coronavirus pandemic tech exploded in size, adding $2.6 trillion in combined market valuation in just a single year. This trend is not about those five tech companies FAAMA (Facebook, Amazon, Apple, Microsoft, Alphabet) alone, but a realignment of the market towards technology capital and value creation. Those stocks have grown because during the pandemic they became more useful, more embedded in people's lives and therefore had greater potential.

Big Tech gets bigger

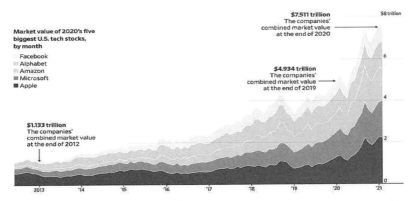

Figure 3: Big tech growth in market capitalization over the last decade. (Source: FactSet [Graphic Kara Dapena / The Wall Street Journal])

The only industry that has remained relatively stable over the last 200 years is finance, but finance is today also being heavily disrupted by technology. In 2020 mobile-based wallets like Alipay, Tencent WeChat Pay, M-Pesa, PayTm, Kakao and others accounted for around $55–56 trillion in total payments value. The entire plastic card industry globally accounted for less than $26 trillion in payments value.

In 20 years, pure-play fintech companies will make up around one third of the total value of the financial services market (currently around $23 trillion annually). More importantly, though, traditional financial services organizations will have all been digitally transformed and be mostly technology-first companies.

Ground transportation will be highly automated and be dominantly electric in nature, shifting core capabilities to battery and storage technology, along with AI-based models for autonomous driving. Ownership of vehicles will have dramatically fallen, and the remaining vehicles will be largely subscription-based or publicly owned infrastructure.

Health tech will rely heavily on gene-therapy, bio-sensor tech that will monitor your health in real-time and AI-based diagnosis capability. Pharmaceuticals will be increasingly personalized based on both your genome and real-time sensor information.

Materials science will be increasingly exotic—with nanotechnology and asteroid mining completely reshaping markets. For mining, while raw materials will still be extracted from the ground, it will be highly automated and driven by sustainability demands and environmental impact considerations. Reuse of raw materials will be an entire discipline in itself.

The future of the largest companies in the world is going to align with technology developments, and stock markets will be dominated by tech-first players and technology growth overall. The ability to extract resources from the ground, to grow food, or to convert raw materials into products will certainly remain elements of the economy, but will not result in corporations that drive market returns and growth. In fact, every leading company of the 21st century will be technology-first and industry specialization-second—so we will have health-tech, energy-tech, autonomous-transport, automated-manufacturing (e.g., gigafactories), algorithmic finance, etc. This is why technology itself will reframe how entrepreneurs and industrialists make their money in the 21st century, and not through more traditional measures like conversion productivity (although that is implied by high levels of automation), monopolization or brand dominance.

Future Dominant Industries

Over the next 20–30 years, which industries or technologies are likely to dominate market growth? We'd argue, as would other commentators, that these might include:

1. **Artificial Intelligence**—whether AGI (Artificial General Intelligence), self-driving vehicles, digital personal assistants,

medical diagnosis or robot delivery vehicles, this is market-maker level change and dominance. AI will pervade every industry within a decade, and while AIs that mimic human interactions may elude us until the 2030s, there are tens of millions of jobs that don't require AGI to be disrupted by automation.

2. **Longevity Treatments and Health-Tech**—Workable cures for old age, or at least the extension of life, will be trillion-dollar entities in their own right. Gene therapy, biotech, medical sensor technology, AI-based diagnosis, are all going to be huge market movers. Life extension will be very profitable and, at least initially, very exclusive and expensive.

3. **Embedded Technologies**—Smartphones will soon be augmented by smart glasses, augmented reality-based personal computers that give you data in a heads-up display. We'll have personal AI assistants embedded in our lives. Smart bank accounts will manage our money. Smart healthcare monitoring will link to AI services, gene-banks and such. Robots will roam the streets, and autonomous drones will populate the skies.

4. **Asteroid Mining**—Don't laugh. With the progress being made in private commercial space vehicles, asteroid mining is definitely a possibility in the 2030s. Asteroid 16 Psyche, one single asteroid mining candidate, is said to be worth $10,000,000,000,000,000,000—or approximately 100,000 times more valuable than the entire world's economy. Mining asteroids like this will accelerate human expansion across the Solar System in the second half of this century.

5. **Climate Mitigation Efforts**—Greening the planet, retooling retail energy generation and distribution, keeping polar ice shelves and glaciers from melting, building city sea-wall defenses that protect from accelerating sea rise, carbon sequestration technology, and removal of pollutants, will all be big employers in the future. While frequently state sponsored, companies that generate unique intellectual property in climate mitigation will be highly valuable.

6. **Global Internet Commerce Platforms**—Starlink and other technology players that provide internet access to the last 10% of humanity, and the global platforms that build entirely new commerce and service layers on the smartphone, smart glasses and smart assistant ecosystems will thrive.

7. **Next-Gen Education**—Education is set for a massive reboot in the 21st century. Akin to the emergence of public education during the Industrial Age, a revolution in education is well overdue and the push for ubiquitous, technology-driven distribution versus the classroom model will be a longer-term winner. Teaching our children adaptability will be critical.

8. **Autonomous Carbon Neutral Transportation**—Tesla is one of the most valuable companies in the world today not because they replaced internal combustion engines that run on gasoline/petroleum with electric motors and lithium-ion batteries, but because they changed the paradigm of automobiles themselves. Autonomous vehicles will lead to subscription services rather than ownership, and electric vehicle (EV) dominance will build entirely new charging and battery technologies and will promote renewable use.

9. **Lab-grown Foods, Robot and Vertical Farming**—Lab-grown meat, fish, chicken—even wine—will dominate sustainable, low-carbon food production, and will shorten the supply chain by putting vertical farms in cities. Robotic farming technology will also make commercial farms highly automated. We will tokenize food on the blockchain for better tracking

10. **Meta-Materials and Nanotechnology**—new materials built on nanotech such as carbon nanotubes, nano-filters used to desalinate saltwater, artificial photosynthesis and better solar photovoltaics are just the start. Over the next 20 years we'll build in-situ resource utilization on Mars and the Moon, create new zero-gravity materials like ZBLan, room temperature superconductors, and the like.

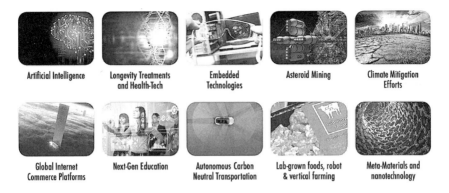

Artificial Intelligence	Longevity Treatments and Health-Tech	Embedded Technologies	Asteroid Mining	Climate Mitigation Efforts
Global Internet Commerce Platforms	Next-Gen Education	Autonomous Carbon Neutral Transportation	Lab-grown foods, robot & vertical farming	Meta-Materials and nanotechnology

Figure 4: The future growth industries of the next 30–50 years.

In any case, with the ironic exception of asteroid mining, almost none of these would be recognizable to the leading companies of the early 1900s. Yes, Ford revolutionized personal transportation, and Tesla is doing the same today, but autonomous, electric vehicles are a clear first principles rethink of the car itself. If you're not driving a car but riding in it in a few years, what do you do in that environment instead? Will we own a vehicle? Will we commute? The rise of new industries leads to many meaningful questions.

The billionaires of tomorrow will fit broadly into three categories: **humanists, innovators** and **disruptors**. The *innovators* are those like Bezos, Ma and Zuckerberg, who can apply new technologies at scale rapidly to expand reach and to build businesses at low distribution costs. The *disruptors* are the Steve Jobs and Elon Musks of the world, who reframe technology in society, change large-scale human behaviour and build businesses that turn industries on their head. The *humanists* are those like Gates, who are engaged in endeavours that enlighten humanity, move us forward as a species, and fight for the causes of the poor and excluded (and are applauded for their resolve and action).

The leading 'industrialists' of the 21st century will employ knowledge workers, those that apply technology innovation to disrupting industries of the 20th century or to accelerating technology adoption more broadly. The richest person on the planet in 2050 will likely be in one of these emerging areas, and not a steel manufacturer, a retailer, a farmer or a coal miner. If we're looking for a trillionaire, it may even be an individual that is as yet unknown but who emerges from AI-based businesses.

Wealth may be AI driven

When it comes to AI—and especially so for the holy grail of AGI, or Artificial General Intelligence[4]—it is clear that while there may be some specialization in various flavours of AI embedded within certain geographies or on certain tech stacks, there won't necessarily be millions of variants of AIs. We can think of the development of AIs like operating systems or app stores. While there is potential for some variations that differentiate in the market, the AIs that will get the best funding will be those garnering the greatest and broadest usage. Those that have the largest active user bases (or interaction depth) will have the best data and learning experiences, and thus will become the most accurate and responsive, further reinforcing their deployment and selection. In this manner, AIs will likely be owned by organizations with the ability to rapidly build their capabilities and deploy them, rather than small tech startups. Who could own the AIs that we use in everyday life? Well, let's first define what we mean by artificial intelligence.

The first mentions of AI occurred in popular literature in the 1850s, beginning with *The Steam Man of the Prairies* by Edward Ellis in 1868, closely followed by Samuel Butler's novel *Erewhon* in 1872. Butler's earlier article, which appeared in the 13 June 1863 edition of *The Press* (a New Zealand-based newspaper), alluded to machines being a type of mechanical life that might someday surpass humans. But even Mary Shelley's *Frankenstein* of 1818 could be regarded as a synthetic being of sorts. The first use of the term robot, or more specifically the Czech term *roboti*, appeared in a 1921 science fiction play written by Karel Čapek titled *Rossumovi Univerzální Roboti* (*Rossum's Universal Robots*). By 1923, Čapek's play had been translated into at least 30 languages. The robots of R.U.R. weren't exactly robots as we think of them today— more like androids or synthetic humans. In more recent fiction they'd be equivalent to Lieutenant Commander Data from *Star Trek*, the androids of the British Sci-Fi series *Humans* or the androids in the computer game

Detroit. However, the term "robot" stuck, and we still use it today. Keep in mind, that in the Czech language robot had the connotation of a serf, a mechanical servant.

It was Alan Turing, however, who first attempted to quantify or define a machine that was capable of simulating reasoning. Turing was the first to propose that "if a human could not distinguish between responses from a machine and a human, the machine could be considered 'intelligent'".

The organizations currently spending the most on artificial intelligence, whether that be the pursuit of general AI or specific capabilities like NLP (Natural Language Processing), are most consistently the tech giants of the West and China—GAFAM/FAANG and BATX[5]. In just the area of NLP, the most common commercial instances of this are AI capabilities embedded in smart home speakers like Amazon Echo or Alexa. Just like the app stores built on top of the Android or iOS operating systems, we don't expect to see dozens of voice-based smart assistants being successful commercially, largely because both the billion dollar development of the AI capabilities and the marketing and distribution of the hardware associated with it requires a massive corporate machine.

In the end, early investors in the tech giants are likely to be the biggest winners in the longer term, or the creators of specific components or skills of AIs that are acquired by these tech giants.

The elimination of poverty meets the widening gap of inequality

In 2018, Oxfam published a report showing that the richest 26 people in the world owned as much as 3.8 billion people, or roughly 50% of the world's population. In 2013, it was 86 of the world's richest people that held the same asset base as half of the world's population. In 2020 the pandemic saw global wealth decrease by US$7.2 trillion. The world's billionaires saw their wealth collectively grow by $3.9 trillion from March through December 2020[6].

Between 2018 and 2020, the billionaire class gained around $2.5 billion a day in new wealth. This resulted in a 19% increase in the wealth of the very richest while the wealth of the poorest half of the world's population fell by 12.8%. In the United States and the UK, where inequality is clearly present, the average billionaire pays a much lower average tax rate than the bottom 10% of tax-paying residents. ProPublica reported in June 2021 that Jeff Bezos paid zero tax personally in 2007, 2011, and 2018[7]. In India it would take an unskilled worker 10,000 years to make what Mukesh Ambani earned in just *a single hour* during the pandemic.

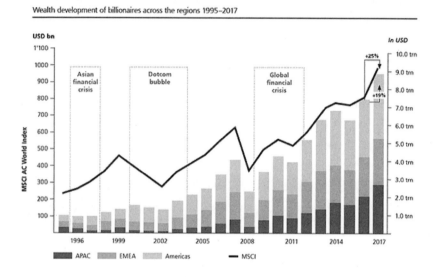

Figure 5: Global growth in billionaires. (Source: UBS Wealth Management and PwC)

The biggest growth most recently in the number of billionaires has been in China, not the United States. It is expected that China will surpass the US in terms of total numbers of billionaires in the next three years.

Proponents of capitalism and efficient markets theory will argue that resource allocation is more efficient in corporations than humans, and more efficiently deployed by billionaires than the poorest elements of society. In September 2018, Elon Musk appeared on the Joe Rogan Experience podcast and stated that billionaires tend to be better at resource allocation than the average person. He argued that they are more driven and have the ability to move the needle more effectively than those who don't have access to large resource pools, nor the experience and ability to administer the conversion of those resources. Musk has said that while he is an engineer and the most significant advances his companies have made have been in areas of innovation, he is also clear that the toughest problems he has had to solve have been in areas of manufacturing and processing to achieve the broadest, most cost-effective adoption of those technologies.

Certainly today, the best method we have in free market societies is directing wealth to the most efficient users of capital, in respect of both output and shareholder returns. However, that doesn't mean that this system is optimally efficient.

Hypothetically, some might argue that a single, well-intentioned populist with exceptional centralized planning could be more effective at resource allocation than multiple agents attempting the same thing. It would appear that the further we decentralize resource allocation, moving away from sensible control mechanisms, the more likely we are to see less efficient mechanisms of value exchange. Perfect competition doesn't necessarily produce perfect resource allocation, because there is a great deal of chaos before a monopoly emerges that is more effective than its competitors.

One big problem with a growing population of billionaires controlling greater portions of national or global wealth is that the incentives and competence of those billionaires doesn't necessarily match up with the wealth they own and the resources they have access to. For example: in the US, the Koch brothers, while being proven generators of wealth over many decades, have arguably not done enough to alleviate poverty, homelessness, food scarcity or climate damage. In fact, they may have contributed to much worsening metrics around elements of those collective concerns.

We're not picking on just billionaires though. The same could be said for trillion-dollar corporations that hold incredible wealth and monopolize the smartest talent and scarcest resources on the planet.

Take Apple Inc, for example. Apple sits on somewhere around $200 billion of cash today—that's more than the entire foreign exchange reserves of the United Kingdom, and multiple times that of Australia. Apple arguably, like Google and Facebook, employs some of the smartest people on the planet. And yet what does that wealth capture do for humanity beyond "the best iPhone ever", a thinner MacBook Air, exotic spatial headphones and touch screens each year? Google has tens of thousands of machine-learning PhDs, data scientists and software engineers; but beyond enabling voice search, or being able to identify a cat in your latest Instagram photo, the actual return to society is not yet significant.

While these organizations and individuals create incredible market wealth and drive economic growth at the GDP level, the level of innovation or improvement we get at a societal level (in terms of basic access to services or financial mobilization, for example) could be argued to be pretty poor. We have been slower to innovate around energy, education, research and development, and medical sciences. How many low-energy electric vehicles were killed off in the past due to patents being acquired by fossil fuel companies, or by starving them of capital? How many cures for cancer have been shelved by big pharma because they couldn't be effectively monetized?

Accumulating wealth, and being able to use it wisely for the purpose of the betterment of humanity, are two very different things. Billionaires and trillion-dollar corporations, therefore, are arguably sources of inefficiency because in capturing huge pools of wealth, they end up restricting access to higher-impact pursuits. These corporations most commonly end up deploying those resources to create more wealth and profits, rather than improving humanity or solving the stickiest problems we face. Obviously, there are exceptions to this, like Musk's vision of a multi-planetary species, carbon-neutral smart cities and EV transportation, or Google's moonshot programs.

The purpose of the market is not to improve humanity *per se*. That much is clear. The market is incentivized to grow, and it does. But can the market be re-tasked or rewarded to shift some of its focus to improving the lot of humankind in general beyond consumption and economic growth?

Can we end poverty, homelessness and food scarcity?

Still the world has attacked the problem of extreme poverty over the last 200 years very effectively. The number of people living in extreme poverty (defined as living on less than $1.90 per day in 2020) has shrunk from 85% of the population in 1800 to just 9.4% today. The real progress only happened in the last 50 years, going from 50% in 1966 to 9.1% in 2017.

Unfortunately, the impact of the pandemic could see the numbers of people in extreme poverty grow to half a billion, the first reversal of progress since the Great Depression. The UN has set the goal of eliminating extreme poverty for the remaining 850 million people globally by 2030.

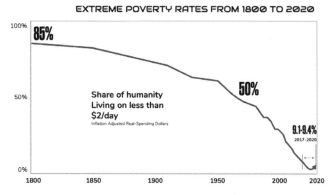

Figure 6: In the last 54 years we've reduced extreme global poverty levels significantly. (Source: WorldBank, OurWorldinData)

China, with the world's strongest economic growth, has seen poverty levels reduced from 98.9 million people in 2012 to virtually zero in 2020. China is proof that a policy to eradicate poverty can be successfully executed. This further implies that a healthy and growing middle class also leads to benefits in the lower-middle income segments of society, as compared to economies where wealth distribution is skewed towards the richest households, such as in the United States, Australia and the United Kingdom.

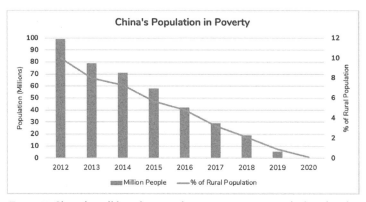

Figure 7: China has all but eliminated extreme poverty over the last decade. (Source: China's National Bureau of Statistics)

While policy is important, many of the key problems facing humanity can be entirely resolved with the application of new, emerging technologies. This is a first in all human history—that we have both the resources and technical ability to solve the toughest problems facing us today.

We'll talk about the technologies that we believe we can use to end homelessness and food scarcity in the following chapters, but the reality is that humanity already has enough resources to ensure every individual on the planet is clothed, housed, fed, educated and live healthy lives above the poverty line—it is simply a question of resource allocation.

Who will be the first trillionaire?

Perhaps the question should be: when will we see the first trillionaire?

Calculating who is likely to be the first trillionaire and when is tricky, but not impossible. Mark Zuckerberg is a relative newcomer to the billionaire ranks, having taken the title at age 23. Bill Gates was the youngest billionaire prior to that, having achieved the status at 31, back in 1987. But most recently, Kylie Jenner beat out Zuck as the youngest billionaire at age 21 in 2017. A decade older when she became a billionaire, but still young, is 31-year-old Whitney Wolfe Herd, the founder of the dating app Bumble, who reached the billionaire milestone in February 2021.

Historically, billionaire status has been attached to either inherited wealth or entrepreneurial success. Jenner's billionaire status comes from her makeup and cosmetics line and the future value of contracts she has negotiated based on her social media influence. Zuckerberg's net worth comes almost exclusively from FaceBook stock value, but wealth for the likes of Bezos, Elon Musk, and Jack Ma comes from multiple companies worth hundreds of billions of dollars. Bill Gates is also more broadly invested, and despite owning a smaller percentage of Microsoft these days, his wealth continues to accumulate. Amazon's value has grown 97,000% since the company's IPO, hence Bezos' wealth has grown in kind.

Historically, billionaires operating family offices with access to investments that most people never get offered, average slightly higher returns than the rest of us. On average, someone like Bill Gates can expect an annual return of somewhere between 10–15%, looking at annual returns for such individuals over the last 50 years. Assuming that their annual income adds to their overall net worth, along with interest payments, if Jeff Bezos has $100 billion today, it will take him around 15 years to build that net worth to $1 trillion purely through income and investments alone. Bezos owns 16% of Amazon today, so if Amazon doubles in value over the next 10 years, you get the compounding effect of the value of Amazon along with his personal investments in cash. Bezos also invested $1 million in Google back in 1998—that investment is worth north of $1 billion today. By 2025, Amazon could be worth over $7 trillion, and if Bezos retains his current investments, then he'll be close to crossing the trillionaire mark sometime near the end of the decade.

Does that sound too soon?

Consider Apple Computers, Inc. Apple was founded in 1976 by Steve Jobs and Steve Wozniak. In 1980 the company went public, selling 4.6 million shares for $22 each. In March 2015, Apple replaced AT&T in the Dow 30, which included the likes of Microsoft, Intel and Cisco at the time. In August 2018 Apple tripped the $1 trillion market valuation level for the first time. It took just two more years for Apple to cross the $2 trillion market valuation level on the 19 August 2020.

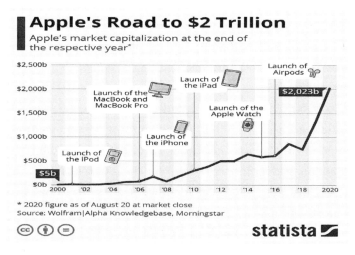

Figure 8: Apple grew from $5 billion to $1 trillion in 17 years, but from $1 trillion to $2 trillion in just two years. (Source: Statista)

Apple's growth is unlikely to continue at its current rate. However, if the company successfully dominates consumer augmented reality device markets as they have the smartphone, and if they expand into other arenas like autonomous vehicles, we could easily see Apple reach a $10 trillion valuation somewhere near the end of this decade. If Steve Jobs' family owned as much equity in Apple as Bezos and Scott own in Amazon, then instead of being worth $10 billion at the time of his death, his family's holdings would be worth nearly $300 billion today.

While Apple and Amazon have grown impressively, many of the fastest-growing companies in the world today are actually in China. Alibaba's, Ant Financial's and Tencent's share prices are growing at the rate of around 18–25% annually. Huawei recently surpassed Apple to become the second-largest smartphone manufacturer in the world, behind Samsung. But what makes Alibaba, Ant Financial and Tencent unique amongst their contemporaries is that their founders own stakes in the businesses similar to US founders, around 10% or more. On the basis of current growth, Alibaba and Tencent both have the potential to become the largest companies in the world by market capitalization over the next decade. Ant Financial is already the world's largest privately-held company[8], worth in excess of $150 billion. Assuming Alibaba and Ant continue their growth path, Jack Ma could surpass Bezos' net worth sometime before 2027— possibly even earlier, if Ant Group IPOs as planned.

Bezos and Ma are clearly candidates, as is potentially some Chinese e-commerce or AI player that is currently emerging in the world's fastest-growing economy. But there is a dark horse in the US economy, one who would have been dismissed just a couple of years ago—Elon Musk. Between 2018–2021, Elon's net worth increased from roughly $25 billion to $200 billion. Mostly this came from a 49% CAGR in Tesla's share price over the last five years, but there are other elements to Musk's net worth that are worth noting.

SpaceX was formed by Musk in 2002. It crossed the $1 billion valuation mark in 2010, and reached $10 billion by 2015. As of October 2020, Morgan Stanley values SpaceX north of $100 billion, with an upper range of $200 billion possible over the next few years. Much of this increase in value is from the launch of Starlink's growing satellite constellation. Starlink alone is estimated to be worth at least $81 billion based on $30–50 billion in annual revenues, and that's before its slated IPO.

If we assess their current holdings and the last five years of growth and extrapolate that out, we have eight possible candidates who could become trillionaires by 2030:

2021 Rank	Individual	Individual Net Worth USD$ Billions (January 2021)	5yr NW growth (CAGR)	Peak Market Cap (Companies in Billions)	Individual Equity (Companies)	CAGR (Company Holdings)	FV Mkt Cap of Company (in Billions)	2030 Value of Holdings (Based on 2020 equity in Billions)
1	Elon Musk	$197	700.81%	$800.55	22%	49.40%	$101,348.92	$22,296.76
2	Jeff Bezos	$182	61.06%	$1,008.00	17%	37.71%	$41,302.75	$7,021.47
3	Bill Gates	$132	34.69%	$813.00	24%	23.39%	$8,243.88	$1,978.53
5	Mark Zuckerberg	$95.6	74.77%	$475.73	28.20%	25.20%	$5,760.35	$1,624.42
8	Larry Page	$81.7	60.51%	$1,210.00	18.50%	15.10%	$5,425.98	$1,003.8
16	Ma Huateng (Pony)	$63.9	296.89%	$440.98	9.70%	38.90%	$20,277.75	$1,966.94
23	Mackenzie Scott	$55.5	54.17%	$1,008.00	4.40%	37.71%	$41,302.75	$1,817.32
25	Jack Ma	$52	38.67%	$472.94	11.70%	51.20%	$71,164.73	$8,326.27

Figure 9: Net worth of the world's richest billionaires by 2030. (Source: Various)

By 2030 it is possible, based on current growth rates in their respective company holdings, that Elon, Bezos and Ma will be the richest three people on the planet, all worth well in excess of US$1 trillion.

All things considered, it seems possible that sometime in the next decade we will see a first trillionaire. Will that trillionaire be South African, American or Chinese? Or will it be someone new—someone with an AI-based technology or longevity treatments that change the world?

We're watching with interest.

Endnotes

1   Bernard Arnault & family recently passed the $201.6 Billion mark to surpass both Musk and Bezos as the world's richest (July 2021)

2   Source: Forbes "The World's Billionaires have gotten $1.9 Trillion richer in 2000", Chase Peterson-Withorn, 16 December 2020: https://www.forbes.com/sites/chasewithorn/2020/12/16/the-worlds-billionaires-have-gotten-19-trillion-richer-in-2020/?sh=606715c87386.

3   Source: World Bank: https://www.worldbank.org/en/news/press-release/2020/10/07/covid-19-to-add-as-many-as-150-million-extreme-poor-by-2021.

4   Also called General AI.

5   GAFAM—Google, Amazon, Facebook, Apple, Microsoft; FAANG—Facebook, Apple, Amazon, Netflix, Google; BATX—Baidu, Alibaba, Tencent, Xiaomi.

6   Source: *The Inequality Virus*, Oxfam International, January 2021 report.

7   Source: ProPublica, "The Secret IRS Files: Trove of Never-Before-Seen Records Reveal How the Wealthiest Avoid Income Tax", July 2021.

8   See Quartz.com "China's Ant Financial raised almost as much money as all US and European fintech firms combined", By John Detrixhe, 30 January 2019: https://qz.com/1537638/ant-financial-raised-almost-as-much-money-in-2018-as-all-fintechs-in-us-and-europe/.

CHAPTER 4

FAREWELL ADAM SMITH?

*"GDP is not a good measure of economic performance,
it's not a good measure of wellbeing."*
—Economist Joseph Stiglitz, Columbia Business School

We need to accept that the economic system that was fit for purpose in the past may not work for the needs of humanity in the future. To better assess this, let's start by looking at past and present economic performance. In doing so we should be able to assess why the present economic system won't continue to work without reform, along with what changes need to be considered for a future-ready economic system to evolve.

We don't want this to devolve into a treatise on economic theory, but let us at least consider the historical elements we need, and those that are increasingly working against us.

In terms of modern economic foundations, we need to look at the basics of economic policy, inflation, interest rates, and how economics in action has charted our course so far; we do this and then we put the growing economic power of China and India into an historical perspective. We will look at productivity and its relationship with inequality. We also need to consider the massive impact that the Global Financial Crisis (GFC) and COVID-19 have had on dislocating the global economy, the way in which those two big problems have been addressed by governments, and the dissonance that is fomenting with regards to how the resulting debt crisis is being managed. We also need to look at forces shaping future economies, such as blockchain, digital currencies, globalization, advances in technology and connectivity, regulation, the rise of the Asia-Pacific region, and the embeddedness of the knowledge economy. That's a lot, but we can't really start rethinking economic systems without considering the good, bad, ugly, and disruptive.

We do not aim in this chapter—or in the book as a whole—to identify all of the problems, and we certainly don't presume to provide all of the solutions. Rather, it is clear that some of the economic ideas that have been widely accepted in the past will no longer work in a digital age and just won't serve a weightless economy. There are some really big issues that need to be addressed more urgently, there are some big gaps that need to be filled, there are some excellent ideas and possible solutions that have already been proposed, and—in the spirit of stimulating a productive dialogue—some other ideas for us to put on the table for debate.

Let's start at the start. Economic policy shapes the world. Employment, investment, money supply, taxation, infrastructure spend, welfare, public debt levels, the machinations and movements of the markets for debt and equity, trade, research and development—and much more—are musicians playing to the tune of the economic conductor waving the policy baton.

Major economic policy decisions taken by governments fall under the umbrella of either fiscal or monetary policy. Various instruments and tonal settings are available for each policy. The idea is that if played correctly, each instrument will enhance prosperity and living standards over the long-term, though this may not happen in the short-term.

Fiscal and monetary policy

Fiscal policy uses government spending and tax revenues to stabilise economic fluctuations and achieve economic policy goals. For many countries during the past 25 years or so, fiscal policy has focused on keeping a lid on debt (not always successfully) and retaining the flexibility to act when necessary. This flexibility was shown during the Global Financial Crisis when many nations responded with a range of fiscal measures, including direct transfer payments to citizens and corporations, along with massive public works programs to stimulate the economy.

However, fiscal policy is not a form of magic that has controllable and predictable outcomes. Too much of it can lead to the public sector crowding out the private sector, upsetting market equilibrium, leading to a mispricing of risk that then leads to suboptimal resource allocation, reduced private investment, and idle assets on the sidelines as the private sector waits for balance to return.

The lag effects of fiscal stimulus, driven by long legislative lead times, a delay in capital deployment, and delays in the start and completion of any public works programs, means that by the time the policy effects are observable and measurable the target may have moved. Too little stimulus and the economy underperforms and possibly stagnates. Too much, and labour demand increases while labour supply stays fixed and inflation balloons. It is a fine balance, and a tricky one to get right. Some of the features of the future economy that we outline in this book are extremely relevant when it comes to policy. For instance, labour supply won't remain fixed if people work for additional years and retire later in life. This would immediately increase labour supply. Increasing participation in the workforce by women has the same effect. It's why things like affordable childcare and education are critical today[1].

Monetary policy uses control of a nation's money supply to try and achieve sustainable economic growth. Central banks use fiat money (currency established by government regulation), to control money supply. Monetary policy can be expansionary or contractionary. During an expansionary phase, interest rates are lowered and the central bank increases money supply by "printing" money[2]. This typically leads to inflation which, if unchecked, can stymie economic growth and lead to significant flow-on problems. In contractionary periods interest rates are increased and assets are sold by the central bank to curtail growth in the money supply. This might slow growth and lead to unemployment, but it acts as a brake on inflation. Most leading economies have adopted an expansionary policy since the GFC and the present outlook is for more of the same.

Ideological arguments often inform the debate as to which policy should be the focus and for how long. Oftentimes this becomes a debate about employment. We would argue that, for several reasons, it is right to focus on jobs, but short-term thinking is being used in making misguided decisions about jobs and employment today. The mission shouldn't be to just create more jobs, but it should be to create jobs that are relevant to the digital, Artificial Intelligence (AI), intangibles-rich economies of tomorrow.

Presently, the combination of large-scale spending, tax reform—which until recently saw tax cuts rather than tax increases being a focus in the US and in many other nations—and the printing of money means that *we are firmly in expansion mode*. Very likely, growing inflation, fiat devaluation, interest rate hikes, and greater unemployment will be the price to pay. (But that's tomorrow, so let's not worry too much about it.)

A note on inflation

So, why is inflation important and what will it mean if inflation spirals upward? Most economists today are in favour of a relatively low and steady rate of inflation and this is reflected in central bank policy around the world, which has tended to aim for annual inflation of around 2%.

Increases in the money supply ("monetary inflation") from money printing generally leads to price inflation, meaning increases in the amount that people pay for things. Right now, a lot of money is being printed and there is the expectation that prices will go up. The question is, by how much?

If inflation increases too quickly—hyperinflation in extreme cases—then people have no incentive to save money because the value of that money in terms of what it can buy is evaporating. So, it is better to buy now at cheaper prices than to lose out later. This can lead to hoarding of goods and cause supply disruptions that further exacerbate the problem. Witness the panic buying of toilet paper and other items considered essential by many during the COVID-19 pandemic as an example of the dysfunction that this behaviour can cause. Now, extrapolate that to food items, clothing, medical supplies, energy, health and beauty products, sports goods, and so on… unhappy days.

If, on the other hand, deflation occurs, then prices fall, wages fall, the real value of debt increases, there is less spending because consumers have less money, more debt and, unless interest rates adjust downward too, more of the money that they do have goes toward servicing that debt. This leads to lower economic growth, which then can lead to the cycle repeating. It is a race to the bottom.

Most economists accept that some inflation is needed, because if prices stagnate or fall then wages are likely to fall as well. Also, controlled inflation helps monetary policy work better because it gives the central bank some power to boost the economy if a recession looms by allowing it to cut interest rates to lower the cost of borrowing money. This encourages businesses and consumers to spend and invest more. And because interest rates are in part calculated based on the inflation rate, if inflation is too low then it doesn't make sense for the central bank to lower interest rates further. In short, a moderate level of inflation gives the central bank the agility that it needs to jump-start economic growth.

Presently, global monetary activity has built up inflation expectations, leading many to seek refuge in consumer prices index-linked securities that hedge for inflation, such as inflation-linked bonds. At the time of writing, the difference between the yields on 10-year US government bonds and their inflation-protected version is 2.43%, the widest it has been in eight years. Inflation is not the only factor to consider when looking at the global economic outlook, but it is an important one, and, right now, the signs don't look good.

The impact of interest rates

As of the time of printing, we've been in a low interest rate environment for years and there is no sign of that changing anytime soon. The US Federal Reserve has said that it won't for now increase interest rates to head off inflation, suggesting that the pandemic means that efforts must be focused on allowing the economy to recover and grow. Money is likely to remain relatively cheap for some time to come, and there is plenty of it. It also means that holding cash isn't going to deliver a return. What to do with surplus cash? Good question.

One unknown is how a generation of earners and investors that has never experienced anything other than a low interest rate environment will react to a change in interest rates. Many of them will have taken on debt to buy homes and other assets—a rational thing to do while the cost of borrowing is low and asset prices are rising. But what happens when interest rates go up and asset prices stop rising, and perhaps even fall?

Much will depend on how quickly each move happens, but we sometimes muse on the fact that during a couple of decades of teaching university students about business, it often surprised us just how many who had not experienced a high interest rate environment were seemingly unconcerned about interest rate changes until they were pressed to consider how their finances would be affected if, say, interest rates moved from 2% to 3%. Nearly always, there were a few students who thought that a 1% increase was no big deal and nothing to worry about. After all, an increase of 1% in anything, even a big thing, wouldn't be disastrous. This thinking was countered each time with the observation that a 1% increase in this context actually represents a 50% increase in the price being paid (for the money). On understanding this, the response from students was very different.

The point here is that these were postgraduate students from varied backgrounds. They represented a good cross-section of society. The students with a financial background, or those working in related fields, understood the interest rate impact before the lesson, but many others didn't. What happens to someone if they have taken on debt at a variable interest rate and interest rates increase? What if rates increased from high 3% to above 16%, the peak rate in the early 1980s? Sadly, they end up, like so many did during the GFC, losing their homes and having to start again. The future economy not only needs to ensure that they will be ok—it needs to be constructed in a way that prevents really bad outcomes from happening.

Economics in action

In terms of government policy and action on managing the economy in the developed world, two main schools of thought have dominated the modern era from the 1920s onward—Keynesian and monetarist (Friedman). There are many other schools, but a comprehensive review of them is unlikely to add much in terms of recommendations regarding the future economy and is beyond the scope of the present work. Each of the two main schools have highly-regarded proponents and has seen success in response to the implementation of their teachings, and each has staunch critics.

During the modern era there has been a Great Depression (1930s), removal of the gold standard, several severe recessions, massive increase in international trade, huge uplifts in productivity, advances in technology, the Global Financial Crisis (GFC), modern trade wars, massive quantatitive easing, massive and recent cases of fiscal stimulus, huge interest rate fluctuations, real world cases of stagflation, hyperinflation, and liquidity traps, and a shift to a weightless economy. Some of these things were theoretical and untested at the time that the core Keynesian and Neoclassical doctrine was divined, and in the case of the weightless economy, that future wasn't even considered.

More modern considerations like Modern Monetary Theory (MMT) have become popular during COVID-19, and could work for Universal Basic Income in stimulating consumption by giving citizens direct access to stimulus payments. When it comes to quantifying a final outcome, the jury is out, but it is clear that the US has been printing money like never before, and many other governments have followed suit. In 2020 alone, the Federal Reserve printed more than 20% of the total US dollars in circulation, and the dollar manufacturing machinery has kept on rolling. As we noted previously, this might be good for now, but what about when the ride stops?

Keynesian doctrine

Keynes believed that government should increase total spending in the economy to support growth. This takes place through actions taken by government directly in the form of taxation and spending—the key elements of fiscal policy—and through the actions of a central bank that buys and sells securities to increase or decrease the money supply—the key elements of monetary policy.

Keynesian policy became popular after the Great Depression of the 1930s. It was regarded as still being market-oriented and private-sector focused, while allowing for government intervention when needed to provide remedies for unemployment. The focus on stimulating aggregate demand was seen as being worker friendly, and Keynes offered a more comprehensive toolkit for managing the economy than previously had been fashionable.

Keynesian thinking led to a massive capital works program in the US and elsewhere, and it inspired tax reform. The interventionist approach advocated by Keynes was in sharp contrast to the laissez-faire capitalism that preceded it and which held that public sector activities should not crowd the market—in short, that government should stay out of it.

The heydays of Keynesianism ran from the mid-1930s until the oil crisis of 1973, when stagflation forced a new approach centred around monetarism. Keynesianism worked well, until one day it simply didn't. Like so many economic schools, it had its time and then unpredicted and/or unprecedented events required new thinking. In recent years, the GFC and the COVID-19 pandemic have witnessed a return to Keynesian deficit financing with the International Monetary Fund (which Keynes had helped establish) and others advocating for those challenges to be met with fiscal stimulus.

Monetarism

Monetarism asserts that the main role of government in economics is to control the money supply via a central bank. Monetarists believe that price stability should be the overarching goal, noting that increasing the money supply too fast will have inflationary effects and contracting it by too much will suck liquidity out of the economy and bring on deflation. This has led to a target rate of inflation being set at around 2% by many central banks.

Monetarism rose in prominence in the late 1950s, but really took hold in the early 1970s as the first oil shock reverberated globally, leading to steep inflation, stock market crashes, and increased unemployment. More restrictive monetary policy was used to fight inflation and featured prominently in the economic policies of the Carter and early Reagan administrations in the US, and in the Thatcher government in the UK.

Many observers now believe monetarism to have failed to deliver on its promise of price stability, noting that the link between money supply and price levels has been overestimated, and that money demand is more volatile and unpredictable than originally thought because it matters very much, in fact, whether the increase in money in the economy takes the form of debt or equity. When debt levels are already high, adding more debt doesn't help much in terms of stimulating real economic activity unless there is government intervention of some kind.

China's economic policy

As the world's second largest economy, but one that is not well understood by many, we feel that China deserves special mention. In fact, we will compare China versus the United States on possible future outcomes regularly throughout the remainder of the book.

China is a communist country that is governed by a single political party. China started down a path of major economic reform in 1978 and today China follows a model of "socialism with Chinese characteristics". This means that the basic principles of Marxism are adapted to suit China's modern-day role as a leading economy, with significant exports across a great many industries.

China has been important to the globalization of many industries and a key connector between East and West. Major policies such as the Belt and Road Initiative (BRI) and the development of the Greater Bay Area, with its unrivalled economic profile and population density, make China a global player that will have a significant impact on shaping the global economy. This is particularly true as more nations become increasingly dependent on China, and as the gravitational pull of China's population on resources and on goods is felt globally. In recent times, the Sino-US trade war has clearly shown the economic importance of China to the rest of the world, and the ongoing status of the trade war has enormous implications for shaping the future economy.

China's economic development is coordinated through Five-Year Plans with the goals of the latest plan, the 14th, being to boost domestic consumption while continuing the rural-to-urban shift that had been featured in previous plans.

We have lived and worked in Greater China for more than 40 years combined, and if only one observation were to be made about China's rise, it is that China thinks and plans for the long-term in a way that few other countries do. This makes it an absolute powerhouse in getting things done, and they're significantly more future proof, in our opinion. While many countries in the West focus on the politics of a 24-hour news cycle (or shorter)—and often prevaricate and dither and end up not deciding on anything in a timely manner because of it—China outlines a long-range strategy across industries and regions, and then commits to that through a range of programs, including co-opting the platforms of China's entrepreneurs.

Monetary policy in China is managed by the People's Bank of China, with the objectives of promoting economic growth, ensuring domestic price stability, and managing the currency exchange rate. China sharply expanded credit during COVID-19, but at time of writing it is starting to ease up on quantitative easing.

At a big picture level, the future of the global economy will likely depend on China more than any other nation. Any major global climate change initiative will need China's support; China is arguably furthest along the path of developing a digital economy and in having a digital currency; China is investing huge sums in AI, genetic research and biotechnology, and quantum information systems; and China's deep economic ties to other countries in Asia and in Africa will give it advantages in accessing many natural resources.

Something else to understand about China's economy is that it was the world's largest in 1820, and for a long period leading up to then. China has, in fact, been the largest economy for extended periods during the past 1,000 years. So in the minds of many Chinese, the nation's re-emergence as the largest economy in the world simply represents restoring the natural order of things, an order that is consistent with history.

What about India?

India holds great promise. For much of the 1800s, the size of India's economy was second only to China. India has the world's second largest population. Only China and India are home to more than 1 billion people. The US, in the number three spot, is way behind with 330 million. China and India combined account for more than 36% of the global population.

Going back to 1980, India and China had economies that were about the same size in terms of GDP—in fact, India was slightly ahead. But over the following 30 years, the Chinese economy grew at an average rate of 10% per annum and India fell well behind. As the world's largest democracy, India had advantages, but its progress stalled as the political system became increasingly complicated, nepotistic, and regional disparities took hold. This led to slow and inefficient decision making, planning cycles that were so long that the relevance of any plans was diminished by the time the plans were approved, widespread corruption, and mountains of red tape that needed to be cut before many activities could occur. India also lagged China in terms of literacy rates, gender equality and healthcare. Decision-making was slow, and its massive human capital potential was underutilized. India's longstanding dispute with Pakistan is another factor that arguably has held back India's economic progress (and Pakistan's too).

Since 2010, India has been playing catch-up, but its GDP is still roughly one-fifth of China's and with India's pandemic response it's likely we will see the gap widen. India has in its favour the fact that some see it as a counterweight to China's dominance and it is likely to benefit from its status as a populous democratic player with proximity to China. It also has a large and experienced industry in serving the offshoring and outsourcing needs of foreign multinationals, particularly in the case of a robust tech sector. The service sector makes up a large share of India's economic output, priming it to be an important part of the future global economy. India will be a force to be reckoned with in the future economy if it gets its act together.

Productivity's key role

The 20th century showed us that increased productivity reduces poverty. The way to close income and wealth gaps is to maximize productivity and then to distribute the resulting gains more evenly. In short, productivity is key to shared prosperity. Provided the right industries are being supported and developed, all economic activity should be geared toward improving productivity, regardless of politics, national boundaries, or policies.

For most advanced economies there was a period in the 1990s during which there were significant improvements in productivity, but since then productivity gains have slowed.

The 1980s and 1990s were a time for major reforms that brought about productivity increases that then tapered off as the benefits were realized and embedded into policies, industries, firms, systems and employee efforts. In many countries, such reforms included deregulation of the financial markets, tax reform, labour reforms, tariff reductions and the privatization of many sectors. Big productivity-driving ideas and reforms fell out of fashion for a time, and, we would argue, are now back on the agenda because of the dual crises of the GFC and the pandemic, in parallel with the emergence of the 21st century digital economy.

Big ideas are needed, particularly as we amble arms outstretched towards higher inflation because of quantitative easing. Provided production costs do not chew up productivity increases from those big ideas, then the unit cost of production will be lower and prices won't need to rise. For this reason, higher production is conducive to lower inflation.

The GFC resulted in a wave of needed regulation and taxation, but negatively impacted productivity in the short-term. For advanced nations, there is the "complacency hypothesis" that suggests low unemployment, relatively high wages, real and propped-up increases in corporate profits, and a shift toward the mindset of "working to live rather than living to work" has dampened productivity efforts.

Another reason for slower productivity gains is that workforce improvements slowed as measured by an increase in the percentage of the workforce with qualifications or specific skills. Some of this can be attributed to the computerization of the workplace, meaning that after learning and incorporating basic computer capability, worker quality and output plateaued at the level needed for job specificity. Obviously broad AI-based automation is about to change that game big time.

Ongoing education, training and development of the workforce is needed, even more so as the rate of change accelerates. In supporting future economies the goal is not to just create jobs to replace those lost by automation. The goal should be to create sustainable well-paid jobs. Ultimately we also need jobs that are both satisfying and meaningful. We need to carefully consider what jobs from the old economy should be saved or kept, what new economy jobs will be created, what industries those jobs are in, where those jobs are located, and what training and development is needed.

Some economists argue that certain core jobs and industries should be maintained at almost any cost. For example, this thinking and the pressure applied around it politically by vested interest groups underpinned the large-scale subsidies of AUD$5 billion given by the Australian government in the years between 2005–2015 in support of the Australian automotive manufacturing sector. At that time, roughly 16,000 workers were employed in that industry. Basic math showed that the government was supporting employment in the sector to the tune of AUD$30,000+ per worker per year. When the industry lost government support, the transition for many workers in that sector was sudden and painful. But let's rewind: what if some forward thinking had been applied and the AUD$5 billion had been spent on retraining those workers to do something else? Something more relevant to Australia's future economy. The vision and long-term thinking were missing, hijacked by a political cycle than runs in three-year terms.

The demographic structure in most economies is also a problem for productivity. An ageing population that retires mid-60s or sooner just isn't sustainable when life expectancy is increasing and retirement incomes are not. In the US, the age for social security eligibility at the full rate has been progressively pushed back so that people born in more recent years have to work for longer. Presently, for anyone born in 1960 or after, the full (normal) retirement age is now 67. But life expectancy for the total US population in 2020 was 77.8 years, meaning that on average workers will live for more than 10 years after they retire[3]. Medical improvements, new drugs, reduced rates of smoking, changes to food regulations, and a host of other factors indicate that the gap between retirement and life expectancy should increase. The OECD notes that if nothing is done to extend working lives, living standards will fall because there will be fewer persons of working age to support more older people. The OECD further notes that a shrinking workforce alongside an ageing population is a matter of social equity across generations because, without reform, it will be future generations who will need to fund the retirement of workers who worked for many years less than the following generation that is supporting them.[4]

As retirees consume their savings, labour costs and other input costs are likely to increase, meaning that the standard of living for retirees falls as their limited dollars are squeezed. One response to this is to introduce higher retirement ages, as has been happening in the US and other places, thereby increasing the labour supply.

If more women join the workforce, and the right jobs are created for workers who are equipped with the skills to do those jobs and they are prepared and incentivized to progressively upskill throughout their careers, and they are prepared to work for more years, then that will be a productivity game-changer. Likewise, if automation and AI create the wealth that is predicted, then caring for the elderly in retirement may become more feasible also.

Productivity reduces inequality

Innovation is at the core of productivity. By increasing productivity across the board we generate innovation, and improve broader wealth distribution. Productivity drives competition, and vice-versa. In the 1970s when Richard's family moved to the Western Suburbs of Sydney his mother was told that it would take six weeks to connect the phone. The monopoly telco provider really didn't care, zero productivity focus. Finally, complaints by many customers about service levels led to reform and the market was eventually opened to competition. Almost instantaneously, six-week service commitments fell to 48 hours; an unforgettable object lesson in productivity and competition. As observed in earlier chapters though, the consolidation of a few leading players in key industries in the US now frustrates productivity as competition eases.

Productivity increases help every level of society and every nation's economic performance is heavily influenced by its productivity. Over the long run, productivity has risen in all countries. Wealthy countries have seen massive GDP increases post industrialization, and especially as they rode the computing and technology wave that really gained momentum in the late 1970s to increase productivity parabolically in some instances. Countries that didn't catch that wave have much lower productivity and are among the poorest. Productivity differentials explain the wide dispersion of per-capita incomes globally, and productivity increases that primarily relate to technology explain the growing gap in wealth inequality in advanced nations.

Basically, if we can nail productivity, then a lot of other good things follow. This is where automation is a double-edged sword. It will create massive productivity gains and wealth, but unless we restructure the distribution of that wealth, inequality will dramatically worsen.

Two big problems that must change the way we see the world

The Global Financial Crisis (GFC) and its aftermath, and the COVID-19 pandemic, have changed the way that people live, work, and think about everything, including the economy. Or they should have.

Prior to the GFC, some of the major forces that are shaping the world were evident. The logic of fiat currencies in the modern era was being questioned and various currencies were periodically attacked, globalization was well underway, connectivity and technology had been improving, cities were becoming increasingly important as connecting nodes, regulatory reform was increasingly being championed, environmental, social, and governance (ESG) and impact investing was being taken seriously, the Asia-Pacific region was ascending, and the knowledge-innovation-creative (KIC) economy was emerging. It certainly was the case that these forces were clear and strong by the time that COVID-19 hit.

But it was the twin GFC and COVID-19 catastrophes, spaced just a few years apart, combined with environmental concerns hitting fever pitch, that rammed home the idea that what worked in the past cannot work in the future. This accelerated reforms, action, and the pace of change, taking the implementation of ideas from a jog to a sprint and erecting clear signposts of what the future economy would be like—rapid cycles of disruptive change, with increasing volatility and uncertainty.

Global Financial Crisis: the first big problem

The 2008 Global Financial Crisis (GFC) exposed structural weaknesses in global financial systems. New types of loans were engineered under policies that increased bricks-and-mortar investment in the US. Agencies like Fannie Mae and Freddie Mac encouraged lenders to extend loans to borrowers who previously would not have been approved; the thinking was that the investments were, to use the popular idiom, "as safe as houses". Mortgage-backed securities were created that received ratings from ratings agencies that shouldn't have been given. The growth in derivatives in the market led to rapid growth in the financial sector but created entire classes of assets that were not properly understood: risk was being mispriced and a massive number of deals were being done that would lead to later problems.

A downturn in the US housing market was the beginning of chaos. The thinking had been that packaging securities in the "right" way would lead to a diversification of risk, but the error in this thinking became clear once the entire asset class started dropping like a lead balloon. Things worsened when it was realized that many large financial institutions had insured their securities with AIG. One large insurer was carrying nearly the entire 'risk' bag, and it was growing heavier by the second. First Bear Stearns collapsed, then Lehman Brothers—soon the entire global financial system looked shaky. Counterparty risk was unmeasurable. Lending dried up. Without credit to grease the economy, the wheels stopped turning and panic took hold. The US dollar weakened, and commodity prices spiked sharply upward. The only way to prevent the global economy from falling apart was for governments to intervene on a scale that had never previously been seen—if not another global depression was almost guaranteed.

The 2008 financial crisis showed the extent to which the global economic system is intertwined and interdependent; how in spite of advanced financial modelling by some super-smart, highly-educated people, risk was catastrophically mispriced. Regulatory systems had been outpaced by changes in the financial sector, most policymakers had a poor handle on the mechanics of modern financial engineering, ratings agencies were blindsided and failed to keep up, workers in the financial sector were compensated in seemingly disproportionate ways to other industries, and their interests and incentives had disconnected from the interests of the rest of the economy.

The GFC resulted in a huge pile of government debt, massive deleveraging by financial institutions that tightened credit globally, killed millions of jobs, and destroyed enormous wealth. It decimated the savings of many people along with their hope for a better future. The system failed, big time.

In the US, massive quantitative easing created huge debts and devalued the US dollar, leaving the US in a far worse position competitively than it had been before the crisis. Bank bailouts by the US Government and others and stimulus packages intended to stave off further tragedy did make things better in the short-term, but worse in the long-run. Tomorrow was being sold to survive today. We didn't ask whether the price being paid was worth it, there simply wasn't time. The crisis demanded immediate action.

This was a predictable crisis, and it should have been a preventable one. We could have been better prepared and, if we had been, the GFC might have been avoided and the global economy would have been better placed to tackle unavoidable future shocks like the pandemic[5].

Other nations fared badly too, with Greece showing early signs of needing life support. As this reality sunk in, the potential impact on European banks and other Eurozone countries led to massive bailout packages being hastily put together by the main EU countries, the European Central Bank, and the IMF to try to maintain order. Thousands of miles away in Japan, the GFC further ripped apart an economy that had been stagnating for years with government debt significantly higher than GDP, a domestic economy that had been unwilling or unable to reform and change, few clear levers of control left to try and fix it, and disagreement as to which levers should be used and to what degree, leaving the nation mired in political paralysis.

The problems created by the GFC led governments to print money with abandon and to try and export their way out of their problems; but, for many, this would only lead to further problems.

If any positive came out of the GFC it is that calls for a new path, a new economic model, started getting louder, and the calls were now more organized and from a vastly increased number of people and institutions. The effects of the GFC were still being felt when the COVID-19 pandemic hit.

The COVID-19 pandemic: the second big problem

COVID-19 has been devastating. It is a massive human tragedy affecting every nation in the world. To give it some scale, total COVID-19 deaths in the US now exceed the total US military deaths for World War I, World War II and the Vietnam War combined.[6]

Looking through an economic lens, the COVID-19 pandemic led to a further round of money printing and government intervention. The increase in government debt has created a crisis of its own and the effects will be felt for many years to come. In the US, several rounds of stimulus, bailouts, and quantitative easing have run up massive debts and have devalued the US dollar substantially. This combination has significantly weakened the US already, but the greatest effects are likely yet to be felt. Drastic action is needed.

At the time of writing, there is hope that some steps are being taken in the right direction with a proposed $3 trillion legislative package that would improve infrastructure, including high-tech infrastructure such as 5G technology, and reduce carbon emissions and economic inequality. But at the time of print it isn't a done deal, and there is a long way to go before the spend, if approved, benefits the economy and employment in the way that is needed. Japan is in an even worse situation, with government debt spiralling and a domestic economy that appears functionally impossible to reform.

Uncertain government finances, a massive public debt burden, and a weakening of the US dollar and other fiat currencies will have a strong impact on employment, aggregate demand, trade, and investment. Uncertainty will probably be a feature of the global economy for a decade or more and with governments printing money and trying to export their way out of difficulty at a time when many exports are COVID-19-constrained, the international competitive environment is likely to be tougher than ever.

Different problems. Same solutions.
The GFC of 2008 and the coronavirus pandemic of 2020 (and onward) were two very different problems.

The GFC was a financial system issue that was so big that it created problems for the global economy. But it didn't affect everyone directly—people who didn't own financial assets did not have as much to lose, while the wealthiest had more to lose, but also a great deal more power and access to policymakers to try and shape a response that would lessen their pain.

Most people who read the news headlines understood that there was a big problem on Wall Street, and they were shocked by many of the stories, by the sheer scale of the bailouts, and by the losses of the banks; but they didn't feel it at a visceral level. The schoolteacher in Boston still taught school to the same students, receiving the same pay check and paid the same bills, and the same applied to the nurse in Idaho, the government clerk in Singapore, and the grocery store worker in Toronto.

For many, the cause and effect link was not made at an individual level between the GFC and the often delayed price and tax increases, wage restraint, credit tightening, and drop in the number of overall jobs, because most people don't join the dots—either because they aren't trained to do so, or because of the time lag between the headline and the impact that they feel, or because they are just struggling to do the best they can with what they have.

The COVID-19 pandemic was very different. It was a crisis that created problems for the global economy, but it also affected every person in every country with lockdowns, travel bans, restrictions on dining out, shopping and other recreational activities, not to mention the fear and uncertainty of what would happen if the individual or someone they cared about got sick.

So, both problems were big and global, but that aside they were very different economic problems. In 2008, the global economy experienced a collapse in aggregate demand, but in 2020 the pandemic was a negative shock to aggregate supply.

The response to solving them by governments and policymakers globally, however, was much the same. Print money, fiscal stimulus, print more money, announce some infrastructure spending (not all bad)—oh, and print some more money. The scale of money printing by the Federal Reserve in response to the GFC made it the largest economic stimulus program ever, anywhere; that is until the pandemic hit. In 2008 the Fed added nearly $4 trillion to the money supply, growing the Federal Reserve's balance sheet from less than $1 trillion to more than four times that post GFC.

Short term, the solution of adopting loose monetary and fiscal policies appears to have worked even better for the pandemic than it did for the GFC. Mind you, the US response to the pandemic is an order of magnitude larger than the response to the GFC, as has been the case for many nations.

On 23 March 2020, the Fed issued a statement indicating that quantitative easing would take place without limit. Within seven weeks the Fed's balance sheet had grown to $7 trillion. The combination of overtime money printing and a fiscal response from the US President(s) and Congress that is likely to exceed $5.3 trillion by way of support for businesses, households, and the wider economy has resulted in a liquidity-driven boom. Following an initial drop, stock markets performed overall very well during the worst of the pandemic, employment numbers are coming back, major banks are paring back earlier estimates of loan defaults, and there is talk of an era of economic success and a rise in consumerism that will rival the "Roaring Twenties" as the effects of the pandemic wane.

The faster than expected improvement in the US economy in 2021 supports the view that the fiscal stimulus has been excessive.

"I can't find any period in history where monetary and fiscal policy was this out of step with the economic circumstances. Not one."[7]
—Leading hedge fund manager and investor Stanley Druckenmiller

There is a growing debate about whether the inevitable inflation will be temporary, reflecting the sharp bounce back from the COVID-19 recession, or persistent, reflecting both demand-pull and cost-push factors. The liquidity provided by central banks has already led to short-run asset inflation, and access to easy credit will stimulate consumer spending as the recovery accelerates.

As of early 2021, inflation has remained below most central banks' annual 2% target for more than a decade. A sharp increase in inflation would create major problems and lead to an era of macro instability. And there is a better than fair chance this might happen—we're already seeing pressure on supply chain in key areas. The stimulus will increase private savings and as COVID-19 restrictions ease there will likely be a spike in demand and inflationary pressure. Our youth talk about spending their "stimies" on Dogecoin, Robinhood crypto and on new smartphones.

A doomsday scenario would see a return to the stagflation of the 1970s. Then, as now, negative supply shocks roiled the global economy. In the 1970s oil shortages wreaked havoc, in the 2020s there are supply bottlenecks across many industries, growing protectionism, exclusionary trade blocs, a simmering trade war between the world's two largest economies, and a fragmentation and shifting of supply chains based on factors other than cost, like supply certainty and the formation of new political alliances—resulting in higher costs of production. Not to mention recent ransomware cyberattacks on key elements of US industry, including gas supplies and meat production.

Continued income and wealth inequality will only serve to complicate policy responses because maintaining social order is going to become increasingly challenging in an inflationary environment where daily essentials are becoming more expensive and the problems of climate change and currency devaluation are deepening. The popular cry of "hedge against the demise of fiat by buying Bitcoin" in the face of $50,000 or more per Bitcoin will only rankle those who don't have more than a few hundred dollars in emergency money[8]. Feeling further alienated and left behind, the divide between the haves and the have nots will intensify. Today, we're kicking that can down the road. That will come back to bite us.

But let's not worry about all that now because short-term things seem to be ok, perhaps medium-term too… but long term? That's the problem. We can't continue to be anchored in the past, mired in the present, and disconnected from the future, because the future is soon upon us.

Consider that a raft of regulatory changes were made in the aftermath of the GFC. Problem solved, right? Well, many of the new regulations were helpful and needed, and they might have prevented the GFC if they had been in place years earlier, but the pandemic is a different problem and many of the changes made to "fix" the GFC have worsened the economic effects of the pandemic.

At the same time, between 2008 and 2020, we've seen massive changes in technology and quantum leaps forward in the life sciences. In terms of overall longer-term implications for the global economy, the blockchain became widespread and understood, and, in parallel, there was the development of Bitcoin and other decentralized digital currencies, the development of decentralized finance in general (DeFi), and along with it "smart contracts" with huge implications for cutting traditional middlemen out of transactions. A reframing of value and information exchange for the digital world.

More recently we've seen the rise of non-fungible tokens (NFTs) that have the potential to shift the way in which creative endeavours are viewed and rewarded. These advances have taken hold and demonstrated that there is an alternative to the "old" system. It is a viable alternative, and in many cases a proven one, and this means that many people are thinking differently and valuing things differently now. Add to this the fact that the very old idea of universal basic income (UBI) has been modernized, studied more closely, and popularized in recent decades and the time is right for change.

In early 2020 the total market value of Bitcoin was around $200 billion; by February 2021 it had eclipsed $1 trillion. Many now view it as a global digital reserve asset, a digital store of value to rival the more traditionally held gold, rather than being a currency for exchange. This is a far cry from a little over 10 years ago, when 10,000 Bitcoins were traded for two pizzas.

Even though we saw massive shifts over the last two decades, policy responses and the thinking of those in charge remains unchanged. Just watch Mark Zuckerberg being interviewed by the US Congress, and this becomes abundantly clear. What about when the next crisis hits, will the thinking and the responses still be stuck in the 20th century? How much will the dollar have devalued? Will fiat money be rejected by more and more players in favour of crypto? Will central bank monetary policy have any effect on decentralized finance markets?

The problem for the past 20 years or so is that policymakers, advisers, many respected business leaders and politicians are planning for tomorrow as if it were yesterday. They are using economic tools and policies that Keynes and Friedman and others developed for a different time. Keynes and those who followed could not have foreseen the weightless economy, so they certainly didn't plan for it!

The major economic forces in play

"Blessed are the young, for they shall inherit the national debt."
—Herbert Hoover

Having dealt with the two most recent shocks, there are some emerging economic forces shaping the global economy that need to be understood and harnessed to ensure future prosperity.

The debt hangover

At a time when public and private debt is growing from an already high baseline (425% of GDP in advanced economies and 356% globally), only a combination of low short- and long-term interest rates can keep debt burdens remotely sustainable. With low interest rates and inflationary pressures mounting, the outlook isn't rosy.

For the US, the ratio of debt-to-GDP was 129% at the end of 2020. This is higher than Greece's debt-to-GDP ratio in 2010, when the International Monetary Fund had to bail Greece out so that it wouldn't collapse economically.[9] If unfunded liabilities are included in the calculation, something required of private companies but not the US government, then debt balloons to more than 500% of GDP. If the US was a corporation, it would be technically insolvent.

Globally, the combination of fiscal stimulus and quantitative easing has ensured that no major nation (and probably none period) will see a reduction in its debt-to-GDP ratio anytime soon. By the end of 2021, Japan's ratio is forecast to be 172%, Italy's 144%, Europe's 106%, India's 99%, Australia's 49%, and so on.

The increase in public debt levels combined with declining productivity, ageing populations, decaying infrastructure, fraying relationships between some of the most powerful nations in the world, the need to fight climate change, and the coming decline in what have been huge tax revenues from fossil fuels[10], is a recipe for disaster that generations to come are going to have to bake into their economies. With housing increasingly unaffordable to mid-career professionals, this will be at crisis levels within a decade. At the present rate, the burden of debt on future generations is likely to be overwhelming and require that a radical new economic model be forged.

The slow demise of fiat

The future of money long-term is not fiat currency that is printed by a central bank. While today the US dollar remains king of the fiat currencies, it is being printed at a rate that is rapidly devaluing its crown. Economists have long recognized that fiat has no intrinsic value, only an agreed value that is based on trust and faith in the Federal Reserve and the US

Government. By many metrics, that trust and faith has been eroding for some time.

Yes, cash still counts. Only fairly recently has cash fallen out of favour. Historically, there weren't other viable options. Credit cards were experimented with in the early 20th century, but didn't become mainstream until the 1970s. In 2018, there were 22.11 billion credit, debit and prepaid cards in circulation worldwide and total spending sat at $35 trillion in 2019[11]. Many industry insiders believe that "peak plastic" has been reached in some markets[12].

"Cash usage has been on the decline for the better part of the last 25 to 30 years, dating back to the early 90s as popularity of debit cards became mainstream ... the trend started to accelerate significantly with the introduction of the iPhone and mobile banking and payment applications that made it easier to manage money and conduct cashless transactions. Now it seems as though we have another [inflection point], and perhaps the most significant, accelerator: a global pandemic."
—Larry Franco, BBVA USA Head of Retail

Credit card companies are increasingly acquiring fintech companies as digital peer-to-peer networks become more popular due to their increasing ease of use, low cost, and low risk of fraud when compared to credit cards and other forms of payment. Players like Stripe, Square, PayPal, Klarna, Affirm and others are rapidly growing, while card usage is flattening. How did we get here?

The removal of the gold standard

It used to be that the value of paper money was directly linked to gold. This system was known as "the gold standard". With gold needed to back up any issue of money, the ability to print money was constrained because the central bank needed to purchase and store gold that was the equivalent of the money they were printing. This helped stymie inflation but led to problems as interest rates increased and as the US government revalued gold in an effort to try and improve the US economy.

As gold reserves flooded into the US and gold production jumped, the Bretton Woods Agreement was struck, making the US dollar the global reserve currency with a fixed rate conversion to gold of US$35 per ounce. Over time, fears that foreign countries would start claiming payment in gold, and various other factors, led to a decoupling of gold and currency. Today, no government uses the gold standard. Britain abandoned it in 1931 and the US finally quit it completely in 1973, both replacing it with fiat money.

We are cashless!

Nowadays, we hardly ever carry cash. Many of us go weeks or months without ever needing it. It can't be used for the online transactions that increasingly dominate our spending, many merchants don't like taking it, even our local barista refuses to hand over our morning hit of caffeine unless we pay by electronic means. There is a sign at the counter that says, "We don't accept cash".

Figure 1: It is increasingly common to see physical stores refusing to accept physical cash (image: Author's own)

Our children don't want cash either. The tried and trusted "clean the kitchen properly and I'll give you five bucks" of earlier times just doesn't work anymore. Our kids want their allowances paid in V-Bucks for Fortnite, Robux for Roblox and perhaps PayPal for online purchases, but certainly not cash.

Pretty soon central banks will stop printing paper currency. In some countries, most notably Finland, Sweden, Singapore and China, the vast majority of payments are already electronic. According to GlobalData, Sweden might be entirely cashless as early as 2023. Payments across Alipay and Tencent WeChat Pay in China totalled $53 trillion (¥347 trillion) in 2019, almost 5x China's GDP[13], and almost twice that of plastic cards globally.

Technology is enabling this revolution. The GFC, the creation of blockchain, the creation of bitcoin and other crypto-assets, the advent of decentralized finance, and the COVID-19 pandemic—all taking place during a 12-year period—has led to a never before seen cycle of change in relation to money.

How much money is there?

Mountains of it, and its growth is snowballing. In the 12 months from February 2020 when the pandemic started to really take hold in the US, the quantity of money in the US economy increased by 26%, or $4 trillion—the largest annual increase since 1943[14]. This compares with an average of just 5.9% since 1982. It is expected that the money supply will grow by a further 12% or so in 2021. If that happens, US dollars in circulation will have increased by nearly 40% in the space of just two years! That can't continue, but taking those dollars out of circulation won't be nearly as trouble free as it was to inject them into the economy. The prospect of the US dollar holding its value in the face of this is dim.

What happens when the value of the dollar drops due to this increased supply and inflation starts to shrink the value of US dollar holdings? People will start to look for alternatives and they look to hedge their risk. There is evidence that many are already doing this by buying gold, and increasingly through bitcoin and other cryptocurrencies.

Why investors turn to gold

Historically, many economists have believed gold to be a solid hedge against inflation and a weakening of fiat currencies. While it is true that gold prices have generally increased against declines in the US dollar, there is data to suggest that holding non-US international stocks might be a better bet. However, the appeal of gold goes well beyond remembrance of the time when its role was to support the US dollar; it holds value as a luxury good, as a component in various technologies, as a reserve asset, and as an investment. It is also scarce, unlike fiat money. It is estimated that approximately 201,000 tonnes of gold have been mined as an all-time total, with roughly two-thirds of this mined since 1950 and an estimated 50,000 tonnes still below ground[15]. That's enough to fill just over 4 Olympic sized swimming pools[16].

Gold prices have increased sharply since the US abandoned the gold standard in the early 1970s, peaking at $2,048 in August 2020 (see Figure 2). The aggregate value of all gold in circulation at today's prices[17] is more than $10 trillion.

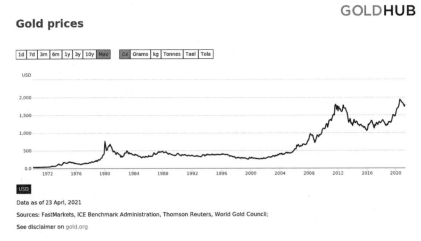

Figure 2: Gold price history since US abandoned the Gold Standard.

It is likely that gold will continue to be a safe haven investment and a hedge against both inflation and a fall in the value of the US dollar. It's scarcity, real use application, and historical and cultural factors all point to long-term increases in the value of gold.

Satoshi's Bitcoin

The ascent of bitcoin since its invention in 2008 and release early in 2009 has been incredible to watch. Its creation has seen perhaps the fastest accumulation of value to a single asset class in history. Bitcoin was designed as a peer-to-peer version of electronic cash. It allows payments to be sent directly from one party to another without going through a financial institution. Bitcoin is globally sovereign, if also decentralized. Transactions can occur without needing any third-party involvement, including government.

The COVID-19 pandemic was bad news for fiat currencies, but it was good for bitcoin. The pandemic forced people to increasingly adopt and use technology, creating a familiarity with online transactions and a trust for them that would otherwise have taken much longer. Perhaps like ourselves you've heard older relatives talk about using online grocery stores for the first time, etc. This rapid shift to online transactions combined with an accelerated trust factor has been good for all cryptocurrencies, bitcoin included. They've become part of the common vernacular.

At time of writing, the total number of bitcoins in circulation is 18.7 million with an aggregate value of roughly $1.2 trillion. So, the total value of bitcoin—created just 12 years ago—is now more than 10% of the value of all gold ever mined. It is also on par with the largest sovereign wealth fund in the world, the Norway Government Pension Fund, with $1.2 trillion in assets[18] (China's sovereign wealth fund is in second place with a little over $1 trillion in assets). To those still sceptical of bitcoin, the scepticism can't be referential to its present overall value. Unlike fiat currency and more like gold, bitcoin is scarce. Once fully mined, a total of 21 million bitcoins will exist. The quantity of bitcoin released into circulation halves every four years. Following that protocol, it is estimated that the final bitcoin won't be mined until 2140[19].

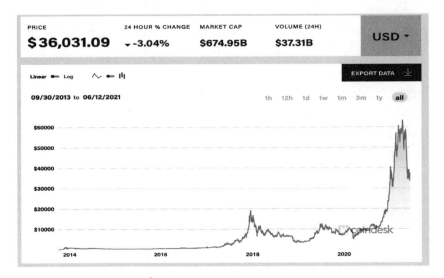

Figure 3: Price of Bitcoin in USD 2012–June 2021.
(Source: CoinDesk)

In more recent times, several companies, including Tesla with $2.5 billion and MicroStrategy with a holding estimated at $3 billion in value, have established themselves as large-scale investors in bitcoin, and highly respected investors including Paul Tudor Jones and Bill Miller have spoken in support of bitcoin as a legitimate store of value. While Tesla has stopped taking bitcoin for Tesla vehicle purchases as of print, they made $101 million profit from their BTC holdings alone in the first quarter of 2021. Cryptocurrency support from institutional investors is growing also, with recent surveys revealing that 60% of institutional investors believe that digital assets have a place in today's portfolio[20]. Bitcoin is a growing in importance as part of the global financial system and many bitcoin analysts expect it to double in value, or more, over the next two years.

A number of cryptocurrency exchanges have been set up to capitalize on the increased value and activity in trading bitcoin and other digital currency assets. Some of these exchanges are publicly listed and are heavily regulated, working closely with regulators to ensure that AML and KYC requirements are closely adhered to, much like banks and other financial institutions. All of this activity is creating a new financial ecosystem that is centred around blockchain technology and is, increasingly, offering more of the financial products and services that traditional providers in the financial services industry offered, just in digital form.

Bitcoin is mainstream, folks. Even Mastercard announced that it will start supporting bitcoin and selected other cryptocurrencies directly on its network[21] and in April 2021, Mastercard announced that it had partnered with cryptocurrency platform Gemini[22] to launch a "first-of-a-kind cryptocurrency rewards credit card"[23].

What does this mean for the 21st century economies of the world? Firstly, bitcoin is likely to outpace gold as a fiat currency hedge, thus leading to economies and consumers alike seeing cryptocurrencies as strong competitors for value exchange. This trust in crypto will no doubt lead to an acceleration of central bank digital currencies that are trying to compete against decentralized systems. We're talking more currency, more value systems. With the emergence of smart contracts and smart money, we'll also start to tie our currencies and investments to themes like carbon neutral operations, sustainability and inclusiveness.

Figure 4: The evolution of Cryptocurrencies over the last decade. (Author's own).

Decentralized finance (DeFi)

Decentralized finance (DeFi) will further liberate the financial system from the ways of the old world. DeFi is an alternative financial infrastructure built using the Ethereum blockchain. Smart contracts are used to create protocols that allow financial exchange to take place in an interoperable and transparent way, more openly than is the case for existing financial services offered by banks and other intermediaries.

DeFi is based on open source protocols and decentralized applications (DApps). Transactions are captured on a public blockchain and completed securely and verifiably in a way that dispenses with the need for much of the traditional financial services infrastructure, such as custodianship, escrow services, or central clearing, and so on. Smart contracts take care of all that.

DeFi is small but growing. Figure 5 shows that the total value of funds locked in DeFi-related smart contracts had been as high as $62.4 billion in April 2021, up from only $833 million a year earlier. That is an astonishing rate of growth that will continue as digital assets and smart contracts become more popular.

Total Value Locked (USD) in DeFi

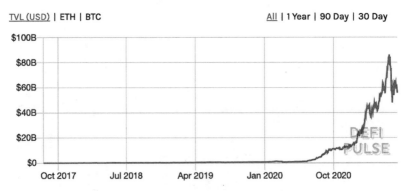

Figure 5: Total Value Locked (USD) in DeFi, one-year chart to 10 June 2021. (Source: defipulse.com)

The DeFi ecosystem has the potential to revolutionize most aspects of the financial system, including the way that debt is issued and managed. Many protocols allow crypto-assets to be used for lending and borrowing.

DeFi means that people can lend directly to others, without needing a bank as an intermediary, and can capture the full amount of the principal and debt repayment. And DeFi loans do not need to identify the parties involved—everything is captured on a decentralized, distributed public ledger. Users remain anonymous while still having the security of there being a permanent record of asset ownership and verification that payments have been made. This is hugely innovative—for the first time in 550 years anyone can essentially become their own bank or lending organization.

DeFi is giving rise to new financial instruments such as flash loans, NFTs (see below), stable coins and atomic swaps, which could not exist without blockchain technology. The innovation being driven by DeFi is creating a more transparent and trusted financial ecosystem, one that will challenge the existing financial industry. Presently, DeFi skews to the younger generation and their use of and trust in smart contracts, and the products available in the DeFi world will dictate how DeFi evolves.

Non-Fungible Tokens (NFTs)

Blockchain technology has also led to the creation of non-fungible tokens. An NFT is a token certifying that a digital asset is unique. The first NFT was created in 2014, but what really sparked people's imagination was when in late February 2021 a LeBron James NBA Top Shot NFT sold for $208,000, which was quickly followed by an NFT based on the artwork by Beeple selling for an astonishing $69.3 million in early March 2021, and shortly thereafter by the sale of the first ever tweet from Twitter co-founder Jack Dorsey for $2.9 million. Those numbers were big enough to capture the attention of pretty much everyone who could read.

The appetite for NFTs is growing and the possibilities for artists, musicians, gamers and others in the creative world to monetize their talents seem limitless. This is because the blockchain technology gives proof of provenance validating the cryptoart as unique, making forgery risk a thing of the past. Content creators can wrap royalty agreements around a sale of their work, giving them a percentage of the transaction value each time the digital asset is sold and creating an income stream, theoretically in perpetuity, for them and their heirs. NFTs are not just a disruption to the way we track assets in the real world, but the way we manage IP. The trademark and patents system of the Industrial Age was never going to be robust enough for the 21st century.

The rapid rise of NFTs is significant to the formation of the future economy in several ways:

1. Puts the focus on digital assets and communicates that they have value, are tradable, are secure, are popular, are encompassing of a wide spectrum of ideas and works, and have greater scale potential and application than many people had thought.

2. NFTs take digital assets beyond owning bitcoin and other cryptocurrencies that are great as a store of value but are not unique—in the case of bitcoin, the one you own will be one of 21 million.

3. NFTs offer bragging rights to the owner. They often relate to digitized entertainment—sport, art, music—and they have a cool factor and confer status.

4. NFTs address the issue of artist's not being paid fully or properly for their work because, if the right smart contract is in place, they will profit from the value of their originality as it appreciates over time. Think of Taylor Swift's IP disputes over her own music.

5. The rise of NFTs has people asking, "What's next for the digital economy?" and wondering what the future possibilities might be.

The move to digital becoming the backbone of the future economy was well underway before NFTs became digital diamonds, but their rise in prominence has cemented the understanding that the future economy is a digital one. Fiat currencies, IP laws dating from the 1800s and asset classes have barely changed since the 1600s.

Digital currency—It's official

The cryptocurrencies we've discussed like bitcoin, ethereum, tether and doge are stateless money today. But they have shown that they have utility and can work to store and transfer value and be used to buy things. This creates a significant problem for governments, because if they lose control over money, then their influence over the economy diminishes. The ability of governments to track and monitor who is doing what in the economy lessens, along with their ability to tax those assets. What choices do governments have? They could ban bitcoin and other stateless currencies.

Well, it's like trying to ban the internet. China banned bitcoin in 2017, then again in 2019. But today around 65% of all bitcoin mining is done in China.

The warning that bitcoin might one day be banned globally was suggested by billionaire hedge fund operator Ray Dalio who noted:

"Every country treasures its monopoly on controlling the supply and demand. They don't want other moneys to be operating or competing, because things can get out of control."[24]

Dalio has also highlighted the benefits and positives of bitcoin. As evidenced by India's government considering whether to outlaw all cryptocurrencies, we see much policy debate on the merits and ability of the government to actually do this, as well as legal action for those in the crypto ecosystem. This led Sidharth Sogani, CEO of cryptocurrency research firm Crebaco Global, to observe: "If the government bans cryptocurrency, it will be impossible to enforce because it does not have the infrastructure to execute the ban."[25]

And imagine for a moment that India banned bitcoin, what then? Well, unless every government bans bitcoin the ban won't work. India will fall behind as other nations find ways to try and regulate and tax bitcoin and benefit from the arbitrage that would arise if some nations ban bitcoin and others do not.

The answer might be in the form of state-sponsored digital currencies, a sort of fiat-equivalent of crypto. Governments clearly see the long-term viability and benefits of digital currencies. China and Russia even suggested that a new digital global currency should be established shortly after the global financial crisis. A global reserve currency "that is disconnected from individual nations and is able to remain stable in the long run, this removing the inherent deficiencies caused by using credit-based national currencies".[26]

This may be why China is seeking to implement its own digital currency. China's 14th Five-Year Plan, approved in March 2021, has further defined the goals of its digital currency. China is an obvious candidate, as a large and globally significant economy, to try to gain a first-mover advantage in respect to state-backed digital money. It processes the largest volume of electronic payments globally, it is home to the largest online payment platforms in the world (Alipay and Tencent WeChat), and has good reason to attempt to break up the US dollar hegemony. A rapidly growing domestic economy and middle class, along with initiatives like the Belt and Road also point to the logic of establishing a central bank digital currency (CBDC). With the backing of the People's Bank of China, this gives it immediate scale and recognition for both local commerce and global trade.

A CBDC has the advantage of being legal tender by virtue of its creation, and payments made using CDBC would not be reversible, as is the case with most other forms of electronic payment. Thus it carries certainty, and is more real-time than other payments options like cross-border wires and transfers. This is likely to set off a new digital currency battle globally with other countries looking to catch China's lead on digital currency. Problems at YES bank in India have raised the idea that India should fast-track the development of a CBDC and led to claims that India's need for a CBDC is even greater than China's[27]. The US too is under considerable pressure to make clear their policy on CBDCs.

Figure 6: China leads the world in development of CBDC. (Author's own).

China and the USA: CBDC part of a bigger strategic move

China, as noted above, is well on its way to the widespread implementation of a CBDC, having started working on it in 2014 and having already trialled it in a dozen cities. China's CBDC will allow the country to gather important economic and transaction data, as well as keep tabs on the activities of its citizens, and it will wrest some control back from some of the large fintech firms whose growth China appears to wish to monitor and keep in check.

China's currency is not easily convertible and as a result the ¥/RMB accounts for just over 10% of international settlements, and, as of Q1 2020 the world's central banks held only $267 billion worth of RMB (compared with $7 trillion held in US dollars)[28]. Thus, the RMB can't yet claim to be a major international currency. But with the development of its CBDC, China aims to channel all cross-border payments made in RMB through a PBOC-controlled settlement system, meaning that those payments would no longer depend on the global bank-run SWIFT network. This would create a payment network to rival the existing trade infrastructure built around the dominant US dollar. This, most critically, would remove the power that the US has in regards to sanctions and export controls[29]. In 2017–2020, we've seen the US impose sanctions on a number of mainland Chinese companies like Huawei, and on a number of officials from China and Hong Kong. It's clear that China has many motives to successfully deploy a CBDC, and that weakening US financial hegemony that is dependent on the US dollar might be a key design element.

The United States is years behind China in the development of a CBDC. Efforts thus far have been circumscribed to Project Hamilton, a joint project by the Federal Reserve and Massachusetts Institute of Technology (MIT), to investigate how a digital currency might work in the US. "Technological advances also offer new possibilities to central banks—including the Fed," Jerome Powell Chairman of the FED said. "While various structures and technologies might be used, a CBDC could be designed for use by the general public."[30] It's clear the US is years away from committing to a CDBC.

> "China is seeking in plain sight to build an alternative international payments system to that of the U.S. dollar, and there's no need to rush to meet this challenge? Nor any thought of actively integrating Bitcoin — a tried and tested decentralized form of 'digital gold' — into the U.S. financial system, rather than treating it as a rather suspect parvenu?"
> —Niall Ferguson, Bloomberg[31]

Why the hesitancy? Perhaps in part it is due the size of China's US debt. In January 2021, China was the second largest holder of US Treasury securities with $1.1 trillion of debt (Japan was in first place with $1.28 trillion). To put that in context, China's position in US Treasury holdings is larger than the value of China's sovereign wealth fund ($1.04 trillion) and is more than 7% of China's total economy, which for 2020 was forecast to comprise more than 17% of the global economy[32]—meaning that China's US debt is around 1.2% of the global economy. That is a huge stake connecting China to the US, and it must concern China that the US dollars it holds will devalue as inflation sets in.

Owning US dollars is a double-edged sword for China. On the one hand, China buying US debt supports the value of the US dollar, making it easier for China to periodically devalue the RMB to make its exports more competitive, thereby strengthening its domestic economy. On the other hand, there are signs that amid a worsening trade war, battles around technology, and negative sentiment both ways regarding the coronavirus pandemic, China is attempting to decouple its economy from the US—and having US dollar reserves works against this objective by increasing China's interdependency with the US rather than reducing it. Perhaps because of this, China has been steadily cutting its holdings of US foreign reserves.

There is often a trade off between currency controls and convertibility. In practical terms, this means without greater freedom in respect to capital controls and RMB conversion, that China would face difficulties overtaking the dominance of the US as a reserve currency. People's Bank of China Governor Yi Gang said that promoting broader use of the yuan will continue alongside the opening of markets. "The regulator's main job is to reduce restrictions on the cross-border use of the currency, and let it take its own course", Yi said[33]. Any Chinese CBDC will face this same balancing act. What freedom is enough to enable trade to shift towards China's eRMB versus more rigorous control? China's CBDC team seems more focused right now on enabling the use of the digital Yuan than overtaking the US dollar from a trade perspective, but that is likely to change based on the Central Bank's commentary.

The prevailing wisdom is that China's US Treasury holdings are sizable enough that it is plausible that China wouldn't do anything to seriously undermine the US dollar. That might be correct, for now. But if China's US debt holdings position shrinks sizeably, that might indicate the start of a very different posture from the Chinese.

Figure 7: Timeline of major fiat impacting events.

1973	US completely abandons the gold standard, replacing it with fiat money
2007–2008	GFC, massive printing of the US dollar, financial stimulus
2008	Blockchain technology invented
2009	First release of bitcoin
2009, March	China and Russia call for a new global currency
2014	First NFT, China starts work on CBDC
2019, December	COVID-19 hits

2020	COVID-19 becomes a global pandemic, massive printing of US dollar, massive stimulus, DeFi ramps up, China completes backend CBDC infrastructure and begins pilot testing in various cities
2021, April	Bitcoin aggregate value exceeds $1.2 trillion
2040–50	Demise of fiat currencies

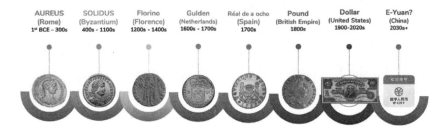

Figure 8: Currency dominance has always been about trade. (Author's own).

We now have a model for what could replace fiat currencies. El Salvador has even chosen bitcoin as its official currency in recent months. We know that won't be using paper money for much longer. Only 20 years ago that was unthinkable to most economists, especially those US-based.

When will the existing paradigm be obsolete? If it is going to happen, it will happen as a result of the winds of technosocialism—more global governance, more collective policies and funding, large-scale global public sector programs around climate adaption and mitigation, the impact of AI, and, of course, decentralization.

The inevitability of globalization

Globalization has been accelerated by improved communication and information systems, lower barriers to trade and investment, new countries opening as markets and production locations, international financial flows, and improved transportation and logistics. This has created new opportunities but also raised competitive pressures. These days, companies are expected to go "global" much earlier in their development at much lower costs than historically was possible.

Digital globalization means that individual countries no longer have to be able to host all of the core activities or infrastructure that a company needs. For developing countries, this is a game changer. The emergence of the internet and global logistics companies has allowed small companies from remote locations to penetrate international markets in previously unfathomable ways.

Globalization has given rise to what has been termed by economists as the "flat world" (no not that one). In this world, the globalization of companies and industries, improved technologies, cheaper and more instant means of communication, and the development of new market economies in locations that previously were peripheral, is resulting in factor price equalization. This means that companies and individuals will no longer earn higher incomes just because they are in a particular location. During the pandemic, some workers even chose to move to a remote overseas location but continue to work for their employers based in their old country residence[34].

Globalization has created more direct competition among companies and workers than ever before. Historically a worker was far better off financially by being moderately skilled but living in an advanced country than being high skilled and living in a relatively backward country. Today, the idea of a flat world means that this is no longer the case. As a result, unskilled workers in developed economies have seen their wages fall in real terms, and companies have chosen to offshore many activities to developing economies.

COVID-19 slowed globalization for a time as a focus on gaps in domestic supply caused supply chains to be reconfigured, but the cutting of new "non traditional" trade deals[35], and China's pursuit of the Belt and Road Initiative, are signs that the trend to globalization will continue to shape future economies.

The cost-benefit of global regulation

Before-COVID-19, SMEs (small and medium-sized enterprises) represented over 90% of the business population, more than 60% of employment, and 55% of GDP in developed economies. SMEs are the backbone of the global economy[36]. There is no economic recovery without an SME recovery, and no economic future can sensibly be imagined without a vibrant SME sector at the heart of it. But trying to understand and then comply with multiple regulatory regimes is becoming harder and more costly for SMEs[37]. This is worsening an already uneven playing field between SMEs and large companies with deeper pockets—regulatory divergence hurts the global economy, but it also creates economic inequality and embeds unfairness into the system.

A common objective should be to understand the impacts of regulation on the development of the economy, as well as on other aspects of society, and to set regulations that are consistent with optimizing the global economy and the development of society for a global citizenry. Smart regulation is essential as it improves confidence and generates economic activity. But regulatory divergence costs the global economy more than USD 780 billion annually[38]. The use of big industry lobby groups in the US to slow down innovation in financial services, crypto and health-tech is clearly evident today, as just one set of examples.

As we embark on large-scale automation and of encoding regulation into the Artificial Intelligence infrastructure of 21st century government, we're going to need to realign policy away from laws that are simply out of date, or that demonstrate biases.

The call for regulatory harmonization has been made for years and great progress has already been in line with increased globalization. In the accounting profession, for example, IFAC and related organizations have made tremendous advances in creating global standards for the accountancy profession, which have paid dividends to the global economy[39]. AI and much more global governance structures around pollution and climate adaptation, will force us to rethink the way we regulate and the way our legal and compliance systems operate in a digital environment.

The Asia-Pacific economic tsunami

Asia has long had the largest share of global population. In 2021, Asia accounted for roughly 60% of the global population. While Asia had the world's leading economies in the days before the Industrial Revolution, its share of world economic output started to decline around 1820. In the 1960s and 1970s Asian economies started to rebuild, but as of 2019,

Asia's share of global GDP was less than 35%, around half its share of the global population.

In 2000, just one Asian economy, Japan, was in the top four (by GDP). Just 20 years later and three Asian countries (China, Japan, and India) are in the top five. In 2010 Asia accounted for just 26% of global GDP, but by 2050 it will exceed half of the world's total economic output[40]. That is a phenomenal shift in a relatively short period of time. By 2050, the Asian Development Bank also projects that Asia's urban population will grow from 1.6 billion to 3 billion and that Asian cities will be leading dynamos of the world economy. The 21st century will be one centred around economic growth in Asia.

China's economy has grown at an average of roughly 10% per year in real terms since the start of its economic reform program in 1979. Even during the Global Financial Crisis, growth slowed only to 8.5%. By 2010, China was the world's second largest economy, leading exporter, leader in terms of international reserves, third leading importer, the leader in inward greenfield foreign direct investment, and the leading producer and market for many industries. The World Bank claims that China will have the world's largest economy at market exchange rates by 2030.

> *"America is rising anew ... We're in a competition with China and other countries to win the 21st Century."*
> **—US President Joe Biden's 100-day address to a joint-session of Congress**

Living and working in the 21st century

The so-called "knowledge economy" was first discussed as such in the late 1960s. This was before the personal computer. The activities performed in business at that time were very different, but the underlying principle is the same—to add value with one's head rather than using hands.

The story so far is that the more advanced economies of the world (i.e. representing about 16% or so of the world's population) have seen a shift from manufacturing as a source of growing wealth to manufacturing as a commodity. From an emphasis on hard infrastructure to an emphasis on soft infrastructure. From competition based on cost control to competition about ideas and intellectual property. From a focus on tangible assets to a focus on intangible assets. Most developing economies are working overtime to try and catch up in respect to infrastructure, training, research and development and innovation.

The rise of economies that can perform manufacturing activities as efficiently and less expensively than advanced economies has forced advanced economies to focus on knowledge, innovation, and creativity as sources of value. The main sources of wealth generation have progressively shifted from production of manufactured assets (like plant and equipment), to the creation of intangible assets (such as software, technology infrastructure, AI and fast-scaling internet start-ups).

Advanced economies are dominated by people who work with their heads rather than their hands. The dominance of the service sector in advanced economies is evidence of this, and the market value of companies

Figure 9: Ten largest companies globally by market capitalization 2021 vs 2011.

	2021			2011		
Rank	Name	Market Cap (USD)	Country	Name	Market Cap (USD)	Country
1	Apple	2.13 Tn	USA	Exxon Mobil	417.16 Bn	USA
2	Saudi Aramco	1.90 Tn	Saudi Arabia	PetroChina	326.19 Bn	China
3	Microsoft	1.85 Tn	USA	Apple	321.07 Bn	USA
4	Amazon	1.64 Tn	USA	ICBC	251.07 Bn	China
5	Alphabet (Google)	1.56 Tn	USA	Petrobras	247.41 Bn	Brazil
6	Facebook	893.22 Bn	USA	BHP Billiton	247.07 Bn	Australia, UK
7	Tencent	768.34 Bn	China	China Construction Bank	232.60 Bn	China
8	Tesla	646.33 Bn	USA	Royal Dutch Shell	226.12 Bn	Netherlands, UK
9	Berkshire Hathaway	645.84 Bn	USA	Chevron	215.78 Bn	USA
10	Alibaba	626.11 Bn	China	Microsoft	213.33 Bn	USA

Sources: https://companiesmarketcap.com
https://en.wikipedia.org/wiki/List_of_public_corporations_by_market_capitalization#2011
https://www.visualcapitalist.com/chart-largest-companies-market-cap-15-years/

that produce intangibles—such as Microsoft and Facebook—has soared past that of traditional manufacturing leaders.

The top 10 largest companies in the world in terms of market capitalization at time of writing are shown in the table below. Alongside them are the top 10 largest companies in the world in terms of market capitalization from 10 years prior.

By 2021 the number of knowledge and intangibles-heavy companies in the top 10 had increased from two to seven if Tesla is excluded[41], with a combined value of approximately $9.46 trillion. Apple and Microsoft alone, saw their combined value increase to $3.98 trillion in just 10 years (a 745% increase in market cap value). This value shift is emblematic of what has been happening in the wider economy. Now, and into the future, the biggest most valuable companies will be technology businesses. This is already the case in many advanced economies, and the pace at which the change is happening is quickening.

This accelerated further during the pandemic. COVID-19 saw the biggest shift in workforce habits since World War II, and it has spurred on numerous disruptive innovations. It has moved us more quickly to remote working, e-learning, telemedicine, new forms of entertainment and new forms of communication, and it has highlighted the readiness of people to change and their ability to do so quickly. The pandemic is proof that the framework of 21st century economics will be fundamentally different, and that consumers and markets alike, will adapt quickly.

Future economics

To get from our present economic systems to the future economy, we will needed to control inflation, reduce debt, increase productivity, adapt to a fast moving digital economy, leverage advances in many new technologies, and adapt to smarter regulation. For now, readers should also understand the core importance of the Asia-Pacific region and particularly the economies of China and India. The realization that the influence of economies like the US and Eurozone players is waning and the rules and engines of growth are changing, too. Above all, finding ways economically to deal with climate change (as the most obvious looming global catastrophe) and Artificial

Intelligence in reframing work and commerce, are critical policy postures lacking today beyond some political rhetoric. We will revisit the economies of the future in Chapter 9—stay tuned.

Endnotes

1 See: "More than 2 million women left the workforce during the pandemic"; ABC News, "Child care at core of women's slow post-pandemic return to work", by Katie Kindelan, 18 May 2021.

2 Often referred to as "quantitative easing".

3 US National Center for Health Statistics (CDC), 2020.

4 Source: https://www.oecd.org/general/ageingsocietiesandtheloomingpensioncrisis.htm.

5 We recognize the debate regarding the cause of the pandemic and whether it too could have been avoided, but let's assume it could not have been and that even if it could then at some point another similar shock would have occurred. The point being that shocks happen and we need to avoid the ones we can avoid and be better prepared globally for the ones that we can't avoid.

6 As at June 2021 the CDC put total deaths in the US from COVID-19 above 600,000. The combined US military deaths for WWI, WWII and the Vietnam War is 580,124. Sources: https://covid.cdc.gov/covid-data-tracker/#datatracker-home; https://en.wikipedia.org/wiki/United_States_military_casualties_of_war#Wars_ranked_by_total_number_of_U.S._military_deaths.

7 https://www.cnbc.com/2021/05/11/stanley-druckenmiller-says-the-fed-is-endangering-the-dollars-global-reserve-status.html.

8 40% of adults in the US, arguably the world's most advanced economy, do not have $400 in emergency money, more than 20% cannot pay monthly bills, and more than 25% skip necessary medical care because they can't afford the cost according to "Report on the Economic Well-Being of U.S. Households in 2017", Board of Governors of the Federal Reserve System, May 2018.

9 Source: American Conservative Movement, "The US government's debt-to-GDP ratio is worse than Greece's before the 2008 crash (and it's about to get worse)", 3 May 2021 https://americanconservativemovement.com/2021/05/03/the-us-governments-debt-to-gdp-ratio-is-worse-than-greeces-before-the-2008-crash-and-its-about-to-get-worse/

10 Historically, energy trades have made up half of the global commodities market, 2020 saw significant declines across the board except for renewables: https://oilprice.com/Latest-Energy-News/World-News/Renewables-Was-Sole-US-Energy-Source-With-Rising-Consumption-In-2020.html.

11 Source: The Paypers: https://thepaypers.com/cards/global-payment-card-expenditure-grew-13-in-2019-to-usd-35-trillion--1245030.

12 https://www.afr.com/technology/credit-cards-on-the-decline-as-visa-says-we-have-passed-peak-plastic-20180628-h11zlw.

13 Source: *South China Morning Post*, "How China's digital currency push can boost fintech and the yuan's global presence", 7 April 2021: https://www.scmp.com/comment/opinion/article/3128475/how-chinas-digital-currency-push-can-boost-fintech-and-yuans-global.

14 *WSJ*, "The Money Boom is Already Here", 21 Feb 2021: https://www.wsj.com/articles/the-money-boom-is-already-here-11613944730.

15 Source: Gold.org: https://www.gold.org/goldhub/data/above-ground-stocks.

16 Extrapolation from Forbes article: https://www.forbes.com/sites/afontevecchia/2010/11/19/how-many-olympic-sized-swimming-pools-can-we-fill-with-billionaire-gold/.

17 Gold Price as at 1 June 2021, $1907.76.

18 See: Top 95 Largest Sovereign Wealth Fund Rankings by Total Assets, SWFI (swfinstitute.org).

19 See: When Will The Last Bitcoin Be Mined? What Will Happen When All Bitcoins Are In Circulation?—Crypto Guide Pro.

20 Source: Fidelity Digital Assets Institutional Investor Survey, 9 June 2020.

21 See Mastercard Newsroom: https://www.mastercard.com/news/perspectives/2021/why-mastercard-is-bringing-crypto-onto-our-network/.

22 A CryptoCurrency Exchange founded by the Winkelvoss Twins.

23 See Mastercard Newsroom: https://www.mastercard.com/news/press/2021/april/gemini-partners-with-mastercard-to-launch-new-crypto-rewards-credit-card-this-summer/.

24 Yahoo Finance, "Dalio sees 'good probability' bitcoin gets outlawed", 24 March 2021: https://finance.yahoo.com/news/ray-dalio-on-bitcoin-and-probability-of-ban-130008375.html.

25 "Bitcoin Could Be 'Outlawed' In Us, Says Founder Of World's Largest Hedge Fund", 25 March 2021, Anthony Cuthbertson.

26 Ahmet Arif Eren, "Economic Issues in Retrospect and Prospect" I, p.268, Ijopec Publication, 2018.

27 *The Economic Times*, "India can use YES Bank debacle to chase China in crypto", 15 March 2020 see: https://economictimes.indiatimes.com/markets/stocks/news/view-india-can-use-yes-bank-debacle-to-chase-china-in-crypto/articleshow/74635079.

28 Source IMF - https://data.imf.org/?sk=E6A5F467-C14B-4AA8-9F6D-5A09EC4E62A4.

29 Source: American Affairs Journal: https://americanaffairsjournal.org/2021/02/carrie-lams-problem-and-ours-chinas-state-backed-digital-currency/.

30 CNBC Money: "The Fed this summer will take another step in developing a digital currency", 20 March 2021.

31 See Bloomberg, "Don't Let China Mint the Money of the Future", Niall Ferguson, 4 April 2021 - https://www.bloomberg.com/opinion/articles/2021-04-04/don-t-let-china-mint-the-digital-currency-of-the-future.

32 China's GDP makes up over 17% of the world economy in 2020: NBS - Global Times.

33 *South China Morning Post*, Nov 2020, "China's central banker remind the global markets that the yuan's challenge to the US dollar is still very much on"

34 See FT.com "The cross-border pitfalls of 'working from anywhere'", 13 April 2021.

35 For instance, the Minister of Economic Development, Trade and Agriculture of Ukraine stated in May 2021 that the conclusion of a bilateral preferential trade deal between Ukraine and the United States can become an effective tool of geoeconomics and would provide increased global competition in trade: https://www.ukrinform.net/rubric-economy/3243941-trade-agreement-with-united-states-to-be-effective-tool-of-geoeconomics-petrashko.html.

36 *World Trade Report 2016: Levelling the trading field for SMEs*, World Trade Organization, 2016.

37 IFAC and Business at OECD, April 2018.

38 *Regulatory Divergence: Costs, Risks and Impacts*, IFAC and Business at OECD, 11 April 2018.

39 See *Nexus 2: The Accountancy Profession—A Global Value Add*, IFAC, 12 November 2015.

40 Asian Development Bank, Asia 2050.

41 We'd argue that Tesla is driven by knowledge and innovation and is an intangibles heavy and clean energy company, but for those who might argue it is a car manufacturer we will leave it out of our aggregation—the point is well made without needing to include Tesla.

CHAPTER 5

OPTIMAL HUMANITY

*"It is difficult to get a man to understand something when
his job depends on not understanding it."*
—Upton Sinclair

Ok, time for the big question. What is the purpose of humanity? Is it to make money or generate wealth? Well, mainstream economics certainly assumes that.

As a species we might reasonably think that our purpose is to learn and evolve. To discover our fullest potential and to make that future possible for humanity as a whole. Should humanity's purpose be dependent on all humans reaching an ascendant state, or is it acceptable if just a small subset of humanity thrives and reaches their potential? Should humanity and our intelligence be protected and valued, or should we simply live in the moment, oblivious to the bigger questions' that life presents?

As we stated at the beginning of the book, there are ranges of possible outcomes for humanity's future, depending on how willing we are to plan, and how inclusive we make the goals of society itself. As we were brainstorming *The Rise of Technosocialism* we started with various white-boarding sessions over multiple days and weeks. We attempted to map humanity's various possible futures, ranging from the dystopian and chaotic to the utopian and organized, the positive to negative. From the most inclusive, planned and objective futures, to the most divisive, chaotic and exclusionary outcomes.

Ultimately, the path we believe is most likely is somewhere in between.

Driven by long-established human behavioural responses, along with our demonstrated collective will to survive once enough risk is apparent. But it also raises very real questions about how outcomes we seek should be framed. Is there a common or collective purpose we should be driving towards? Or is our future likely to be determined by the most successful model of governance and planning that emerges over time?

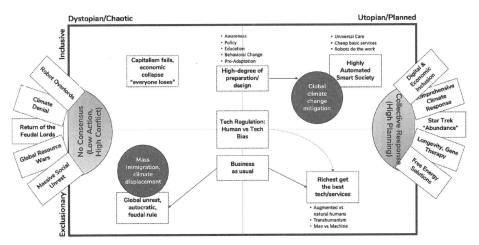

Figure 1: Possible socio-economic futures.

We debated a host of ways to map out humanity's possible future. One theme emerged over all others: **How much focus and purpose we apply to our future through collective planning and action.**

We have identified four possible outcomes which are Luddistan, Technosocialism, Neo-Feudalism and Failedistan (see matrix on the opposite page).

We've already identified key risks that are destabilizing society, and those that represent existential risks in terms of humanity's future and broad economic uncertainty. Our ability to respond to those risks in a coordinated fashion requires at a minimum some sort of broad commitment on behalf of the planet and humanity. The only one of these four scenarios that we believe tends to support greater equality is a planned, broadly equitable society based on leveraging technology to mitigate climate risks and inequality at the same time (Technosocialism).

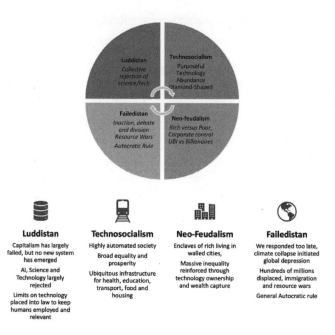

Luddistan

Capitalism has largely failed, but no new system has emerged

AI, Science and Technology largely rejected

Limits on technology placed into law to keep humans employed and relevant

Technosocialism

Highly automated society

Broad equality and prosperity

Ubiquitous infrastructure for health, education, transport, food and housing

Neo-Feudalism

Enclaves of rich living in walled cities,

Massive inequality reinforced through technology ownership and wealth capture

Failedistan

We responded too late, climate collapse initiated global depression

Hundreds of millions displaced, immigration and resource wars

General Autocratic rule

Figure 2: Possible scenario-based system outcomes.
(Source: author's own)

Macro outcomes for society are often complicated by where we sit ideologically on a spectrum of fundamental issues, such as:

1. **Human Rights**
   (a) Everyone is born equal with certain unalienable rights that should ultimately equate to a basic quality of life, *versus*:
   (b) Survival of the fittest! If people are born with certain advantages or work harder than others, so be it

2. **Economics/Money/Wealth**
   (a) The free market is the best organizing principle humanity has found for growing prosperity, *versus:*
   (b) Money is an arbitrary abstraction and if we got rid of it we would be better off

3. **Nationalism vs Globalism (Individualism vs Collectivism)**
   (a) The future depends on us finding some way for humanity to come together, *versus:*

(b) God has blessed my tribe and the ground upon which we were born, anyone else is a competitor for resources

4. **Ethics**

(a) Humanity should strive to live to certain moral standards, *versus:*

(b) Let God sort it out in this life or the next

In any normal period of human history, we would simply debate these differing points of view via academic process or across the political aisle. But all that takes time, and rarely results in consensus. For example, the world's philosophers have been debating God's existence since time began and we've hardly reached consensus on that issue alone.

Our beliefs and assumptions shape personal decision-making processes, but in the aggregate, they act to form public policy. History has shown us that our decisions as a whole can have unforeseen consequences, whether positive or destructive. Today, however, we face a cascading series of potentially extreme macro-level risks that require much more concerted effort and intent to solve. Debate is no longer efficient—just look at our world's climate. We need a minimum level of action to achieve a better than disastrous outcome, one where humanity itself is at risk of extinction.

Assuming the fundamental risks we identified earlier are real, the issues remaining are: Can we agree on collective strategies to mitigate said risks, or are we set to debate the merits of action until the worst consequences are largely inescapable? Can we act collectively for the good of humanity, or will we try to gain some advantage for our tribe that makes consensus improbable? Are we willing to make potential sacrifices now in order to ensure the success of future generations, or will we just keep kicking that can down the road?

Greek philosopher Aristotle (384–322 BCE) argued that the purpose of man was to find individual purpose, and if done virtuously that purpose would result in happiness both for the individual and community. It's as good a central premise for humanity as any other philosophy.

Logically speaking, anything that gives the human species long-term survivability and enhances our chances of individual success, prosperity, and longevity, should be celebrated, right? What sort of species would want anything but an optimal future for our offspring decades and centuries from now? If that is a sensible underlying assumption, then it's clear that today we are not on an optimal path.

> *"Articulating [a] clear understanding of the end toward which a person's life aims, Aristotle states that each human being should use his abilities to their fullest potential and should obtain happiness and enjoyment through the exercise of their realized capacities. He contends that human achievements are animated by purpose and autonomy and that people should take pride in being excellent at what they do. According to Aristotle, human beings have a natural desire and capacity to know and understand the truth, to pursue moral excellence, and to instantiate their ideals in the world through action."*
> —Dr. Edward Younkins, *author of* Capitalism and Commerce

We're not really attempting to answer the question of what the purpose of humanity is—that would be arrogant, or at best futile. However, it is a reasonable assumption that the less we reach consensus on the actions required for a healthy and prosperous future, the worse off humanity will be. When it comes to climate in particular, the less we do now the greater the potential negative impact for our grandchildren.

The failures of short-term focus

When viewed in the context of the ability of humanity to thrive, we might even ask whether capitalism is the best model we can possibly think of to advance the species as a whole? There's a reasonable argument that capitalism does incentivize innovation and the advancement of industries, economies and markets at a macro level; but is it the best model we can conceptualize, full stop? Probably not. The likelihood that capitalism in its current form will still be here 1,000 years from now is extremely remote, given the lessons of history. But capitalism has another more fundamental flaw.

Capitalism, by its nature, could be argued to aid in more frequent exclusionary outcomes when it comes to various economic classes, especially lower- and middle-income households. Capitalism also generally leads to shorter-term economic objectives, less rational decision making in respect to social policy, and more divisive political ideologies. Politics and 24/7 mainstream news networks also keep us focused on extremely short-term issues, and not tackling longer-term problem resolution.

When humans attempt collective long-term planning, they generally stumble at making tradeoffs over multiple generations, or coming together with collective purpose, unless forced to by extreme circumstances—like warfare, famine, pandemics, etc. These are the rare exceptions, and even then, they don't necessarily lead to consensus.

The Great Wall of China is an example of long-term commitment to outcomes. Other examples include research into the human genome, the Apollo space program, the national highway project in the US, and the multi-century commitment made by our ancestors to build the great cathedrals in Europe. Throughout most of human history, though, we've clustered policy and government spending on much shorter-term objectives.

This natural tendency to focus on shorter-term outcomes creates a whole host of problems individually, too. We focus on the next paycheck, without saving adequately for potential emergencies or retirement. We eat too much processed food, find ourselves constantly too busy to get to the gym, and inevitably our biometrics head in the wrong direction. We buy shares in companies that perform well one quarter and sell them off when they miss targets the next quarter. We buy a new iPhone when it's released or buy a Starbucks coffee every day, instead of putting funds away for our children's future education.

When individual rights curtail the collective good

Most philosophers argue that individuality and individual expression are vital to human fulfillment. We want independence and the right to our own determination; but it is also true that the more personal freedoms we demand, the more likely it is that we will infringe on the freedoms of others. Individual rights frequently are a trade off against the rights of those around us.

The gap between the US Bill of Rights and the UN Declaration of Human Rights is one framed around the specificity of citizen's rights as respects "life, liberty, and happiness". The US constitution leaves these matters for open debate, whereas the UN argues these basic rights are a function of a modern economy. The only way for these rights in the US to become explicit is for the broader citizenry to force the government's hand - these aren't guaranteed constitutionally, and hence remain elusive.

The chances of successfully committing suicide with a gun are almost 20 times greater than by a drug overdose or using a knife. In Britain, there are about 0.06 violent gun deaths per 100,000 residents; in the US it is roughly 4.43 violent gun deaths per 100,000. Even accounting for population differences, gun homicide rates in the United States are 73 times higher than the UK. The US accounts for roughly 4% of the world's population, but half of all privately-owned guns in the world. While some have argued that guns prevent crime, around 30 carefully framed studies have shown this assertion to be wrong—more guns are linked to more crime[1].

Enshrining the right of individuals to own a gun in the US has restricted the right of society at large to feel safe and protected from the risk of mass shootings, from gun violence at home and a much higher probability of death from suicide. If you believe that giving up your gun(s) is a threat to your "personal rights", then as a gun owner you'll fight on principle to retain that right—rather than consider evidence that shows that society would be better off if gun ownership was more tightly managed.

Gun ownership is a clear example of the prioritization of individual versus collective rights. The US Constitution clearly emphasizes individual rights and aligns closely with Jefferson's own philosophical beliefs. In most developed nations collective rights are guaranteed first, and individual rights second. The US is fairly unique in this way.

> *"We hold these truths to be self-evident, that all men are created equal, that they are endowed by their Creator with certain unalienable Rights, that among these are Life, Liberty and the pursuit of Happiness. That to secure these rights, Governments are instituted among Men, deriving their just powers from the consent of the governed."*
> **—US Declaration of Independence**

The Declaration of Independence makes clear that individuals have "unalienable rights", and indicates that government is permitted only by the collective consent of the governed. At the time of framing the constitution, however, these unalienable rights were only extended to white landowning citizens—women and African-Americans were explicitly excluded. The constitution, therefore, does not really address the collective needs of Americans as a whole because it assumes that if you guarantee the rights of individuals, they will ultimately be free to choose the best course. But the pursuit of "life, liberty and happiness" in society is, in actuality, governed by reasonable tradeoffs between the rights of the individual and the collective good, and the constitution as a document didn't really attempt to resolve this.

The rights guaranteed constitutionally in the United States include the right of free speech, free press, free worship, trial by jury, and freedom from unreasonable searches and seizures. The United Nations articulates humanity's broader view of rights and includes a much wider range of socio-economic principles, including the right to work, the right for equal pay for the same work (regardless of race or gender), a basic right to healthcare and education access, social security support and so forth.

> *"Everyone has the right to a standard of living adequate for the health and well-being of himself and of his family, including food, clothing, housing and medical care and necessary social services, and the right to security in the event*

of unemployment, sickness, disability, widowhood, old age or other lack of livelihood in circumstances beyond his control."
—United Nations, Universal Declaration of Human Rights

The US Constitution only explicitly guarantees those tradeoffs when the people stand up and force the government's hand. Not as a basic function of the Bill of Rights.

We see a similar problem emphasized during the COVID-19 pandemic across the world. Where individuals asserted their rights to ignore lockdown rules and refused to wear a mask as an "individual right", without considering those around them and that the consequences of their decision might further spread the virus. The only guaranteed way to manage such pandemics more efficiently in the future is submitting to the science, restricting our own actions and acting in line with health authorities' recommendations, specifically around vaccines, masks and social distancing. But that also assumes that we educate our citizens adequately so they can trust the core science.

We could reach the required level of crowd immunity[2] that will safely protect everyone if we adhere to these constraints. But continued emphasis on the individual rights of a citizen to reject a vaccine or not wear a mask has thus far simply lead to avoidable deaths. When we act as individuals without a collective framework, it leads to suboptimal outcomes for the species.

"Some feel face coverings infringe on their freedom of choice—but if more wear them, we'll have MORE freedom to go out."
—US Surgeon General Jerome Adams

Ultimately, the fact that some saw masks as an infringement on their personal rights, versus those that saw it as a personal responsibility to protect their fellow citizens and neighbours, will be one more element of the ongoing debate about acting for the collective good.

Freedom of speech is a fundamental right in most societies today. However, social media gave us a vehicle for enabling more radical examples of free speech to flourish, such as the QAnon phenomenon, hate groups, extremism and fake news. While it is admirable to prevent a government from stamping out criticism, if we don't also have the tools to stop groups that incite violence against the community, collectively we're arguably worse off. Certainly, the January 2021 riots and incursion of the US Capitol Building are proof of that conundrum. The free speech that led to the Capitol riots included assertions about the 2020 presidential election that were proven unfounded by US courts. Free speech contributed to a sense of purpose and anger among those present at the breach of the Capitol building, a group that might otherwise have remained a fringe element.

The cohesion of society, politically and economically, obviously requires such tradeoffs. The more inclusive a society becomes, the less wealth the rich accumulate. The more protection governments assert, the less freedoms we individually have. The more security we put in place, the more intrusive policing becomes. The more we rely on technology and science, the more we see historical norms and employment undermined. The more gender and sexual equality we allow, the more we see long-held social and religious traditions diluted.

Trying to predict how we organize and shape society in the coming decades appears to centre around two key philosophical issues: the first is the belief that we can affect future outcomes through action, the second is whether we should sacrifice the rights of individuals to further the purpose of the collective.

We would argue that humanity must commit to giving our children the best chance of solving these problems, but only if we see the future happiness and needs of our children as being equivalent to our self-actualization and needs in the present.

Tribalism and self-interest erode freedoms

For many, life appears to be an Olympic sporting event where the results are binary—you win, someone else loses. The world is full of marketing messages emphasizing privilege, rewards, premium service and exclusivity.

At the foundation of many of the debates and fractious discourse today is how ready and willing many are to divide our world into "us" and "them". This mentality leads to us viewing outcomes as win or lose for our tribe or our group—even though such groupings are often arbitrary and amorphous.

If an individual thinks that it is acceptable for others to lose in order that their tribe wins, how can society ever expect that same individual to make sacrifices for the benefit of their great-grandchildren that they'll likely never meet, let alone someone born on the other side of the planet?

This tendency to act in our own self-interest over the interest of others, or indeed contrary to the common good of society, is known in economics as the "tragedy of the commons". In 1833, the British economist William Forster Lloyd, during his residency at Oxford University, wrote a pamphlet describing the problem of overgrazing on public land by individual cattle herders. He described the situation where an individual herder might abuse the common good by putting more than his fair share of cattle out to graze. While the individual herder could benefit and might rationally justify his decision based on individual economic benefit, the common good would be negatively affected, resulting in detrimental effects overall to the society around him.

In 1968, ecologist Garrett Hardin published a paper in the journal *Science* entitled "The Tragedy of the Commons" in which he explored Lloyd's social dilemma in respect to the use of Earth's natural resources and unrestricted population growth. Hardin argued that when humans act individually without taking into account their relationship with society as a whole, that a Malthusian catastrophe would occur where we deplete all of Earth's resources. Hardin held that conscience was an ineffective method of policing the commons, because selfish people would always win out against those who were more altruistic. Hardin concluded his paper by stating, "freedom is the recognition of necessity"[3], that for humanity to survive its own extinction, it needed to recognize the planet and its resources in its entirety as commons. Hardin argued that management of these finite, shared resources was the only way to "preserve and nurture other and more precious freedoms".

Some have argued that the data for climate science is inconclusive, that we can't possibly know what impact Artificial Intelligence will have on society longer term, or that if computers do reach that level of consciousness someday, there's no evidence they will be malevolent. Devil's advocate positions like this lead to a cycle of debate over action.

Belief sets that are self-destructive

We cluster our "beliefs" not only around philosophies and religious frameworks, but around political ideologies or moral arguments that fit our tribal narratives. We discussed earlier in the book that economic uncertainty has been a core driver of recent political debate, but psychologists point to other issues that allow people to consistently vote against their own interests, and those of society more broadly.

In the United States, farmers living in dominantly GOP "red" states voted strongly in favour of Donald Trump in both the 2016 and 2020 presidential elections, despite losing $14.4 billion[4] from the former president's trade war with China. White women voted 62% in favour of Trump, not Hillary, in the 2016 election, even though Trump had allegedly abused women and was caught on tape talking about abusing women[5], and then, after aligning with the Republicans, flipped his views on abortion and women's rights. The poorest white Americans also voted for Trump, knowing that the GOP had historically supported tax cuts on the wealthy, were opposed to universal healthcare and free education, worked frequently to frame laws around lobby group interests, and that statistically a GOP president was much more likely to increase debt and slow GDP growth than a democratic president[6].

Why would farmers, women and the poorest rust-belt Americans vote for a president who would statistically be likely to make their life worse off? When 9 out of 10 economists predicted Britain would likely suffer from an economic downturn as a result of Brexit, why would the segments suffering the highest levels of unemployment in the UK vote for a Brexit that could result in fewer jobs overall and slower economic growth?

Writing in *Psychology Today*, Dr. Bobby Azarian identified 14 key psychological phenomena influencing those who voted against their best interests, including the Dunning-Kruger Effect, how fear acts to stimulate conservative brains, terror management theory, relative deprivation, racism, and so forth. But three key themes continually emerged in the psychology of conservative voters: firstly, the perceived moral framing of a candidate's policy platform; secondly, the dominance of patriarchal family units; and finally, fear and uncertainty around future prosperity.

A 2008 study published in the US National Library of Medicine (NLM) suggests that political leanings may, in fact, have a biological basis around how the brain responds to threats.

"Individuals with measurably lower physical sensitivities to sudden noises and threatening visual images were more likely to support foreign aid, liberal immigration policies, pacifism, and gun control, whereas individuals displaying measurably higher physiological reactions to those same stimuli were more likely to favor defense spending, capital punishment, patriotism, and the Iraq War. Thus, the degree to which individuals are physiologically responsive to threat appears to indicate the degree to which they advocate policies that protect the existing social structure from both external and internal threats."
—"Political attitudes vary with physiological traits", Department of Political Science, University of Nebraska-Lincoln, 19 September 2008

Fear, uncertainty and doubt

Scientific evidence suggests that conservative brains process fear and uncertainty differently from more progressive voters. The "fear" they experience is much more visceral, processed as something more personal and individual—they react to threats as if they are directed at their immediate family. Whereas liberal brains are more likely to process threats as impacting society collectively, and be less sensitive to individual threats.

Logically the more fearful you are, the more likely you are to vote for more conservative policies, and inevitably for a maintenance of the status quo as a recognizable form of safety or normalcy. It's one of the reasons politicians often evoke immigration, crime or threats to the economy— it's as old as time itself. The bad guys are over there, and they are coming to take your jobs, money, land, etc. Let's build a wall to keep them out. This worked for the Chinese when the Great Wall was built (700 BCE to 1644 CE), and it arguably worked for Boris Johnson with Brexit and Trump with his MAGA mission.

A research study conducted in 2003 by Stanford University in cooperation with University of California and University of Maryland showed that collective social cognition might be linked to uncertainty, systemic instability, complexity, threat and fear responses. They provided evidence that following 9/11, America collectively experienced a conservative shift, where voters became more strongly aligned with George W. Bush and direct military action. These results provide support for the motivated social cognition model of conservatism[7] over predictions derived from historical models. Jost's study (2003) was conducted in five countries across 22 separate tests, which all appear to confirm the hypothesis that fear and uncertainty fuel more conservative viewpoints. According to 2011 research at the National Centre for Biotechnology Information, those who most often identify as conservative have been shown to have larger and more active right amygdala[8], the area of the brain that processes fear. A clearly biological response for some, but also environmental—flight or fight response.

Seen in this context, 9/11, the Global Financial Crisis, Al Qaeda, and even the pandemic are viable stressors that contributed to the rise of more conservative policy and voters who are more likely to support Trump and Brexit. Populist movements clearly employed fear and threat triggers around unfettered immigration, declining economic conditions, crime and healthcare issues. Ironically, on the left more progressive voters responded to those same pressures by giving support to the likes of Bernie Sanders, and what might be considered more socialist leanings. While it might appear logical that positive messages should be the best at uniting human behaviour toward a collective response, recent outcomes suggest that a fear and threat response may be far more effective at uniting people.

Is fear the human element that might eventually relegate us toward a more dystopian, divided and chaotic set of futures? Will the chaos of climate change, gross inequality, increasingly worrisome pandemics, protests and conflict, push us finally to unite in the common cause of a better future? Are we even capable of acting in the best interests of each other while such threats exist?

Is species-wide collective action really viable?

Let's ask that question a different way. What will it take to motivate humanity to embark on a multi-generational effort to undo the damage to our climate? For historical precedents, when we review the largest projects and human efforts throughout history, what typically motivated large groups of people to act?

- The Great Pyramid of Giza, Egypt: completed at a period of great collective wealth, made in worship of Egyptian pharaohs who became divine beings at death. In today's dollars, about $5–10 billion (Religion).
- The Great Wall of China: built over a 2,000-year period to protect from invasion by tribes from Inner Asia. In today's dollars about $65–90 billion (Fear of Invasion, Sign of Wealth).
- The Apollo Program, United States: built to land Americans on the Moon first. In today's money, $146 billion ($24.5 in 1970s) (Fear of Russian dominance, Exploration).
- The Human Genome Project: a 19-year global effort (1984–2003) to encode the first human genome. About $5 billion in today's dollars (Healthcare Innovation, Longevity, Scientific Endeavour).
- International Space Station: 100 launches, 100 space walks, over 1,000 cubic metres of pressurized volume, a mass of nearly one million pounds (420,000 kgs), the size of a football field, traveling at 28,000 km/h. In excess of $150 billion spent over 20 years with 18 participating countries (Space Research, Global Science Development).

- Panama Canal: built between 1881 and 1914 with first French and then American expertise. About $9.5 billion in today's dollars (Commerce).
- The US Interstate Highway System: 62 years of effort, 46,000 miles of Interstate Highways for the purpose of national commerce and defense, at the cost of approximately $450–500 billion in today's dollars (Infrastructure, Defense).

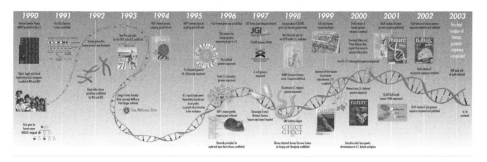

Figure 3: Large scale human collaboration produces advances faster than market innovation alone. (Source: Human Genome Project).

What could motivate mankind to come together on the issues of climate change and AI?

Coordinating a climate response is probably closer to the Interstate Highway System, Apollo, the Panama Canal, or the Hoover Dam—a mix of infrastructure development, global scientific/technical response, and national defense strategy. This would involve retooling energy production to renewable sources on distributed grids, new energy storage techniques, geo-engineering, coastal sea rise defense systems, rehousing displaced populations, ensuring adequate food production and supply chain improvements, environmental sustainability in manufacturing, reducing carbon production, carbon sequestration and capture, mass recycling, and such.

By 2025, the world's coal plants will be too expensive to run when compared to renewable energy schemes across the globe. In fact, for 75% of today's existing coal plants it would still be cheaper today to replace them entirely than keep them running[9], and that includes the cost of building new, unsubsidized solar and wind farms from scratch. It's significantly cheaper just to replace all coal infrastructure, rather than keep it running. Replacing all fossil fuel generation facilities across the planet will take decades of infrastructure development, but the cost savings to governments over the next two decades alone would easily pay for it. By 2030 it is estimated that at least $255 billion of stranded coal infrastructure will be lying dormant. We should be doing this just based on cost, regardless of carbon emissions; but honestly, cost shouldn't even be the primary consideration.

Can we all agree that a renewable future is necessary, regardless of capital considerations? Obviously, there are those that still disagree (coal companies and their paid-for politicians), but the vast majority of humanity, once they understand that this will create tens of millions of new jobs, save millions of lives a year, and that it's cheaper and cleaner, should get behind it, right?

Whenever you talk about climate mitigation, we are immediately talking about quadrillions of dollars of capital being deployed over the next 50–100 years. It is almost too big a problem to wrap our head around, and certainly committing to that sort of spend over that length of time is political suicide in a conventional sense. That's why we suggest forgiving national debt in order to raise the required amount of funding that could make a real difference to the climate fight over the next 30–50 years. But clearly if we're going to respond to climate risk, we're going to have to think bigger. A lot bigger.

The development of Artificial Intelligence is more similar to the Human Genome Project or the International Space Station—a global, joint scientific endeavour, but with clearer and nearer-term commercial applications. The application of AI will involve investment in research and development, foundational ethics for the operation of AI, risk mitigation strategies for impact to employment, national core technology infrastructure, public policy and strategies, etc.

Is global consensus on climate and AI realistically possible?

Figure 4: How did Earthrise change mankind's view of the world?
(Source: NASA Apollo 8 Archives)

On Christmas Eve in 1968, astronaut Bill Anders quickly snapped a shot of the Earth as it came into view as CSM-103[10] came around the dark side of the moon and back into radio contact with mission control in Houston, Texas. Little did Anders and the Apollo 8 crew realize the profound effect that this picture—"Earthrise"—would have on humanity, as this was the first time we had ever seen our home planet from this perspective. In fact, nature photographer Galen Rowell declared it "the most influential environmental photograph ever taken"[11].

Less than two years later on 22 April 1970, the world celebrated the very first Earth Day, stimulated by an environmental awakening. Greenpeace was also founded in the years following "Earthrise". On 2 December 1970, the Environmental Protection Agency was founded after President Nixon had made it a priority in July of that year. A plethora of anti-pollution, anti-nuclear and environmental protests and activism emerged collectively soon after the "Earthrise" image appeared. Seeing our home from space appears to have profoundly changed us, but more importantly it showed we could have an awareness of a higher collective purpose beyond pure capitalism and economic outcomes.

What could be our Earthrise moment in the 21st century? Will it be the flooding of New York City, the inundation of Bangladesh or the Maldives? Will it be 10 straight years of the worst wildfires and bushfires the world has seen? Will it be the collapse of the world's insurance industry from an overwhelming concentration of natural disasters? The problem for humanity is that all of these events would be too late for us to mitigate the worst effects of climate change because they will already be happening.

Globally coordinated efforts to address our future will need us to identify the forces that impede progress, that reinforce inequality and exclusion, that have sponsored policies that benefit small groups of the population, and that have led to fear being used as a political lever. We must seek to eliminate those systemic elements of economics and politics, which by their nature have demonstrated long-term negative impact on humanity's advancement, wellbeing, and prosperity. Ironically, that may include elements of capitalism and democracy where they conflict with humanity's prosperity.

The resistance: battling vested interests

While fear may have crippled our collective political responses over the last few years, there are other groups that have consistently worked hard to restrict progress in certain key areas, spending trillions of dollars collectively to prevent erosion of their commercial interests through regulation, funded policy positions and restrictions on reform. These include:

- **Military-Industrial Complex (MIC):** $1.7–1.8 trillion dollars annually is spent on so-called "defence" spending. Military Keynesianism is the economic theory that war makes the economy grow, although the US' recent forays into the Middle East have worked against this theorem.
- **Big Tobacco, Alcohol, Guns and Pharma:** numerous exposés and whistleblowers have uncovered massive commercial efforts by these groups to influence legislation, subsidies, funding and research to benefit their industries. One classic example is the

NRA sponsoring laws in the US Congress that prevent research on gun violence and deaths[12].

- **Processed Food and Junk Food:** Americans alone spend over $200 billion on junk food annually. By 2022 the world will consume more than $700 billion worth of junk food. Eating junk food just once a week is linked to high rates of obesity, while twice a week is associated with type 2 diabetes, death from coronary heart disease, depression, increased risk of cancer, and cognitive issues[13].

- **Commercial Fishing Industry:** more than 50% of plastics that are polluting the world's oceans come from commercial fishing vessels and not from single use plastic straws or bags[14]. Freshwater fish have declined by 76% in less than 50 years[15]. Seafood may have disappeared entirely by 2048 at current commercial fishing rates.

- **Big Oil, Coal, Gas and Energy:** recent investigations have shown that companies like Exxon, Royal Dutch Shell and BP knew about climate change in the 1970s and worked continuously to frustrate climate research and spent billions on funding climate misinformation. The largest five stock market listed oil and gas companies spend nearly $200 million a year lobbying governments to delay, control or block policies to tackle climate change.

- **Australian Coal** (specifically): recent bushfires in Australia have drawn attention to the significant political contributions by Big Coal to the conservative liberal party and their representatives. While small-fry compared with lobby groups in the US, the combined spend on promoting pro-coal and anti-renewables sentiment is in the tens of millions.

- **Big Tech:** in 2018 Facebook, Amazon, Microsoft, Apple and Google spent almost $70 million on lobbying politicians in Washington. Comcast and AT&T spent another $30 million plus. Since 2005, Big Tech has spent over half a billion dollars trying to influence Capitol Hill[16].

- **General Politics and Lobbying Spend:** the US spent $14
  billion on advertising and media buy alone for the 2020
  presidential elections, more than twice that of the 2016
  elections[17]. In the US just 0.26% of the population contributes
  68% of political contributions thanks to cases like Citizens
  United. But maybe the real winner (or loser) in the 2020
  election was Facebook.

These are just some obvious examples. We could include planned obsolescence in consumer products, the war on drugs, charity and NGO fraud, the illuminati…

The point is that all the money we spend on vested interests is more than enough to pay for any corrective action that we might need to take on climate change, or investments we could make in respect to universal healthcare, housing the homeless, or mitigating job losses from automation. Cut defence spending in half, and we can eliminate student debt in the US within two years. Eliminate spending on election advertising and we'd be able to provide free education for approximately one million students annually. Redirect fossil fuel lobbying efforts back into renewable jobs and we could create tens of thousands of new jobs annually. Reduce the $1.3–2.6 billion it takes to get a new drug to market, and we could make prescription medicine a fraction of current costs.

What will it take to cause enough pushback against such vested interests to produce real systemic reform?

Climate change will spur on some reallocation of funds here. Take the US military. A 2019 report by the Royal Geographical Society[18] determined that the US military is one of the largest polluters in history. The Army, Navy and Marines collectively consume more fuel and emit more climate-changing gases than most medium-sized countries. If the US military were a country, its fuel usage alone ($8.7 billion and 98,268,950 barrels of oil annually) would make it the 47th largest emitter of greenhouse gases in the world. This is the reason why the United States government insisted on military emission exemptions from the Kyoto Agreement in 1997.

Downsizing the US military and greater build-up in cybersecurity forces rather than physical armed forces may help here.

Political system change? That's a tougher one. Maybe real-time elections based on real, collectively-curated policy, where spend is dramatically restricted per candidate to produce a more level playing field? Perhaps policy and laws that are voted on by citizens or where the vote is shared equally between legislators and constituents? We're getting closer to technology that might make all of that possible—we'll talk about that in our closing chapter. At some point in the future, resource constraints might find us using AI to optimize resource allocation, and this may evolve into policies that correlate much more closely with societal outcomes from a health, wealth and access perspective, as those will be the most logical to code into regulatory artificial intelligence.

Perhaps the real policy changes will be forced by collective fear over social collapse expressed through virtual revolution and powered by a new sort of capitalism that prioritizes technologies that work for the collective good, rather than just delivering returns to satisfy the capital markets.

Endnotes

1   *Scientific American*, "More Guns Do Not Stop More Crimes, Evidence Show", 1 October 2017.

2   Crowd immunity, not herd immunity. We are not bovines.

3   Also known as Hegel's Maxim.

4   Source: National Bureau of Economic Research, "The Impact of Retaliatory Tariffs on Agricultural and Food Trade", May 2020.

5   See NYTimes transcript of Access Hollywood exposé: https://www.nytimes.com/2016/10/08/us/donald-trump-tape-transcript.html.

6   Source: US Congress, Joint Economic Committee, "The Economy Under Democratic vs. Republican Presidents", June 2016.

7   Jost, J. T., Glaser, J., Kruglanski, A. W., & Sulloway, F. J. (2003). "Political conservatism as motivated social cognition". *Psychological Bulletin*, 129, 339–375.

8   Ryota Kanai, Tom Feilden, Colin Firth, Geraint Rees (2011). "Political Orientations Are Correlated with Brain Structure in Young Adults", 10.1016/j.cub.2011.03.017.

9   Source: CarbonTracker, "Powering down coal: Navigating the economic and financial risks in the last years of coal power", November 2018.

10   Apollo 8 Astronauts didn't name their Command Service Module like later flights, and hence it was just known as the CSM for Apollo 8.

11   "That Photograph", an interview with Galen Rowell, renowned nature photographer by the Australian Broadcasting Commission in 1999.

12   "How The NRA Worked To Stifle Gun Violence Research", NPR Radio, 5 April 2018.

13   "Fast Food Pattern and Cardiometabolic Disorders: A Review of Current Studies", Bahadoran Mirmiran and Azizi, January 2016, National Institutes of Health scientific journal.

14   Source: Sea Shepherd Organization: https://www.seashepherdglobal.org/latest-news/marine-debris-plastic-fishing-gear/.

15   Source: National Geographic: https://www.nationalgeographic.com/animals/article/migratory-freshwater-fish-decline-globally.

16   Source: Fox Business, "Big tech has spent $582M lobbying Congress. Here's where that money went", Megan Henney, 26 July 2019.

17   Source: London School of Economics, "The 2020 election was the most expensive in history, but campaign spending does not always lead to success", by William Horncastle, 27 November 2020.

18   Source: Royal Geographical Society, "Hidden carbon costs of the "everywhere war": Logistics, geopolitical ecology, and the carbon boot-print of the US military", Belcher et al, 19 June 2019.

CHAPTER 6

GIVE ME YOUR TIRED, POOR, HUDDLED MASSES...

"Remember always, that all of us, and you and I especially, are descended from immigrants and revolutionists."
—Franklin D. Roosevelt

When Donald Trump announced he was running for President of the United States, his first order of business was to attack immigrants, those perceived to be entering through the southern US/Mexico border. "When Mexico sends its people, they're not sending their best," Trump quipped. "They're sending people that have lots of problems, and they're bringing those problems with us [sic]. They're bringing drugs, they're bringing crime, they're rapists." Basically, "immigrants are coming over the border to kill you", is the way Dara Lind from Vice characterized Trump's Republican primary speech repertoire.

Why immigration is an essential economic component

Trump went on throughout the GOP[1] primaries to claim that migrants were taking American jobs and contributing to economic displacement, "while some we assume are good people". Aware of the growing unease of poor and middle-class Americans at the performance of the US economy, and the lasting memories of the Global Financial Crisis, Trump was providing a target for the crowd's anger and emotion. It worked.

Figure 1: Nigel Farage was unapologetically anti-immigrant throughout the Brexit debate. (Credit: SkyNews)

Boris Johnson and Nigel Farage tapped into the same concerns and uncertainties with the UK Brexit movement, ultimately resulting in a successful leave vote, along with Boris Johnson's election as prime minister. Farage is quoted as saying he believes that Brexit would not have "got over the line" if it hadn't been for the single issue of immigration. Farage argued that mass immigration was hopelessly out of control and set to get worse if Britain remained in the EU.

Figure 2: Farage tweeting about immigration being "out of control". (Source: Twitter @Nigel_Farage, 26 May 2016)

In reality, economies like the US and UK are going to desperately need immigration to continue to deliver economic growth over the next few decades. Additionally, climate change and automation are going to create massive migration potential, so there's no scenario where we can just ignore the problem of migration by attempting to shut down borders, as the sheer numbers make such a proposition all but impossible.

Ultimately there are three key reasons why immigration will become a hotly contested economic issue over the next two decades:

1. **Economic stimulus**: historically, immigration has created demonstrable economic growth that is not possible without it.
2. **Declining birth rates**: China will reach peak population in the middle of this decade, the United States as early as 2040 (depending on unemployment rates which impact birth rates); dozens of countries' populations are already shrinking, and this will have a clear impact on economic growth.
3. **Climate eco-refugee explosion**: in 2017, 68.5m² people were already forcibly displaced due to climate change and regional conflict—by 2050 that figure could be as high as one billion people. As discussed in previous chapters, it is estimated that 300 million alone will be displaced due to coastal sea rise.

Economic effects of immigration

Over the last 40 years, across multiple political positions, multiple studies have found that immigration is of significant benefit economically in countries like the US[3]. Recent research from the American Enterprise Institute (AEI), show that the facts around immigration don't support Trump or UKIP's stated position on immigration. AEI research shows that between 1990 and 2014, US economic growth would have been 15% lower without the benefits brought by inbound migration. The same research shows that the UK would have had economic growth 20% lower without immigration, and across the EU 20–30 points lower. Given the 2008 GFC, almost all the post-crisis gains in the US economy could be linked to migration.

The New American Economy Research Fund found in 2019 that 45% of Fortune 500 companies listed on the US stock exchange were founded by immigrants, which was up from 40% in 2011. Those companies generated $16.1 trillion in revenue in 2018 alone, out of $20.58 trillion of GDP. Clearly the US economy would look very different without immigration.

IMMIGRANT FOUNDERS/CO-FOUNDERS

Elon Musk
Tesla/SpaceX

Sergey Brin/Larry Page
Google

Pierre Omidyar
EBay

Garrett Camp
Uber

Eduardo Savarin
Facebook

IMMIGRANT CEOs

Sundar Pichai
Google

Satya Nadella
Microsoft

James Quincy
Coca Cola

Dara Khosrowshahi
Uber

Ajay Banga
MasterCard

Figure 3: Some of the immigrants who led US Fortune 500 companies over recent years. (Source: Various)

The total impact of immigration on the economy has long been a topic of heated debate. Most economists agree that immigration is a strong net positive for growth. President Trump's alma mater the Penn Wharton School of Economics published a long-term impact research study in June of 2016. The research showed that immigration (both legal and illegal) leads to more innovation, a better educated workforce, greater occupational specialization, better matching of skills with jobs, and higher overall economic productivity. Immigration also has a net positive effect on combined federal, state and local budgets. However, the research found that in regions with large populations of less educated, low-income immigrants, there was a negative impact to the overall cost of public services, especially in respect to education. The research concluded that illegal immigration into the US was marginally positive for the economy, whereas for legal migration the results were extremely positive.

The OECD[4] found significant economic benefits attached to immigration in research across multiple countries, using multiple different metrics and methods. Specifically, they found:

- **Immigration tends to boost the share of the population employed**: increases in the share of workers in an economy or population growth through immigration both lead to per capita income increases. Specialization driven by highly-educated immigrants further stimulates per capita income.
- **Immigrants' contribution to value added often exceeds their population share**: the contributions of foreign-born workers to the economy ranges from about 1% of GDP (in Ghana) to almost 19% (in Côte d'Ivoire). In most of the countries studied, estimates in GDP value added correlate with the share of foreign-born workers in employment.
- **Econometric models illustrate the contribution of foreign-born workers to GDP**: even where immigrants were low-skilled workers, their introduction into the economy increased productivity of low-skilled workers overall. In countries like South Africa, high-skilled foreign-born workers raised GDP per capita by 2.2% and low-skilled raised GDP by 2.8%.
- **Immigrants improve culture and productivity at an individual firm level**: firms employing immigrants tend to grow faster than firms that don't, and they also deploy less capital per employee. Immigrants tend to transfer new skills to existing employees, thus improving productivity. Overall productivity increases by as much as 25% where just 5% of the workforce are immigrants.

The US was founded on immigration-based growth and, arguably, the economic stimulus that immigration delivers. Across the 19th and 20th centuries, immigration rates meant that 15% of the US population was consistently foreign born. While there was a significant decline in immigration for the first half of the 20th century, today the US has returned to that long-term 15% mark.

Foreign-Born Share of U.S. Population, 1850–2017

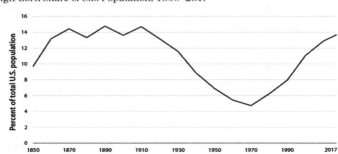

Source: American Community Survey (ACS; U.S. Census Bureau [Census] 2000–17); authors' calculations; Gibson and Jung 2006.
Note: Data for 1850–1990 are from Gibson and Jung 2006; data for 2000–17 are from the ACS. U.S. citizens born abroad and residents of U.S. territories are considered native-born.

THE HAMILTON PROJECT
BROOKINGS

Figure 4: Foreign-born share of US population (1850–today).
(Source: Brookings Institute)

Almost half (49%) of the Australian population today is an immigrant or the child of an immigrant. Australia's projected population will be 38 million by 2050 and migration will be contributing $1.6 trillion to Australia's GDP by then. Overall, by 2050, each individual migrant will on average be contributing approximately 10% more to Australia's economy than native residents.

In 2018, the Australian Treasury produced an extensive report on the long-term impact of immigration on the nation. The report found overwhelmingly that immigration had a strong positive impact on society. In fact, the conservative Liberal government's attack on immigration policy during the 2019 election was done fully cognizant of the fact that reducing immigration into Australia would cost the national budget billions of dollars and reduce job growth significantly.

"Migrants deliver an economic dividend for Australia due to current policy settings which favor migrants of working age who have skills to contribute to the economy. This leads to higher rates of workforce participation and likely productivity benefits. This, in turn, increases Australia's

*GDP and GDP per person, with positive flow-on effects for
living standards…migrants are likely to contribute more
to tax revenue than they claim in social services or other
government support."*
**—"Shaping a Nation: Population Growth and Immigration
over time", Australian Treasury, Department of Home Affairs**

Researchers at Wharton went on to show that immigrants "are at the forefront of innovation and ingenuity in the United States, accounting for a disproportionately high share of patent filings, science and technology graduates, and senior positions at top venture capital-funded firms." In respect to taxation, the study found that immigration "improves the government's fiscal situation, as many immigrants pay more in taxes over a lifetime than they consume in government services". Unemployment impact? While immigrants increase supply of labour into the market, the purchase and construction of homes and consumption of food, TVs, electronics and other goods and services by immigrants, lead to broad job growth for native-born Americans. Immigration has consistently been one of the US' secret weapons in creating the world's most innovative country. For 50 years we've been hearing about the great American Dream and why immigrants should move to the US. This Hollywood-based messaging has been extremely successful in attracting some of the best talent from around the world.

Trump's executive order (EO) of 22 June 2020 restricting individuals seeking to enter the country on non-immigrant working visas barred the entrance of nearly 200,000 foreign workers and their dependents. The Brookings Institute calculated the immediate effect of this was a $100 billion loss of value for the US' top companies, who rely on H1-B visa workers. Gallup reported in July 2020 that 34% of Americans would like to see immigration increased, whereas 28% would like to see it decreased; 77% of Americans say that immigration is good for the country. Despite this, immigration fell from an average of close to one million per year to around 200,000 in 2019, largely because of the administration's changes to immigration policy and the bad PR associated with the administration's public position on immigration. If immigration was to stay at this reduced level over the next decade, Moody Analytics estimates that US GDP would be negatively impacted by at least $1 trillion.

Despite the attacks made by politicians continuously over decades, it is likely that immigration will become a key feature and even one where economies compete to attract immigrants in the 21st century.

Competitive immigration

Let's return to the biggest elephant in the room. Demonizing immigrants is a problem. Firstly, immigration of any sort is generally positive economically and creates jobs, rather than destroying them. But the big issue is that climate change and automation are both set to trigger migration of massive numbers of people. As birth rates decrease in developed economies, one strategy that will undoubtedly emerge is of countries competing to attract skilled immigrants, especially those versed in technology like AI, engineering, renewable energy and climate response competencies. Why? Because the world will force developed nations to take higher levels of immigrants due to climate displacement, and everyone will scramble to get the highest numbers of skilled workers possible in that allocation.

If we don't have a plan to accommodate 20–50 times the level of global migration we see today, we're going to run into significant issues. What are these issues?

- **More porous borders**: as the level of migrants and refugees seeking new homes increases, border integrity failures also increase. We have not seen long-term success with shutting borders as it pertains to decreases in illegal immigration. Despite President Trump's harsh immigration policies during his term in office, the US still saw an increase from 43.7 million to 45 million foreign-born persons living in the country (approximately a 3% increase). This is even with more Mexicans leaving the US than those arriving (due primarily to improving economic growth in Mexico).

- **International pressure on refugee programs**: it is likely that the United Nations, WHO and OECD will increasingly push the largest historical CO_2 producers to take a greater share of climate refugees, perhaps even correlated with total carbon output. In fact, Bill Gates is already a big proponent of this.

- **Resource conflict**: with masses of climate refugees and global attempts to mitigate the largest humanitarian crisis in history, resource allocation to refugee efforts could take as much as 15% of global GDP. With decreases in arable farmland due to global warming, increasing food scarcity, and more demands on resources, we can expect significant conflict.

Bottom line. By 2050 we will need global and national planning for how to absorb immigrants.

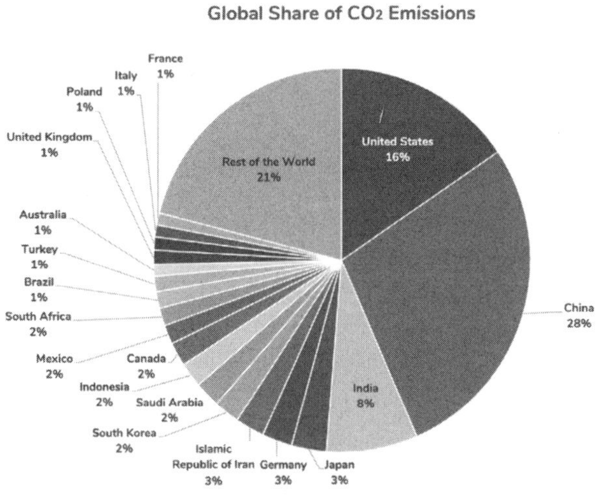

Global Share of CO2 Emissions

Figure 5: Could international bodies pressure the largest polluters to take a greater share of climate refugees? (Source: Union of Concerned Scientists UCS, 2020[5])

Rethinking education for the 21st century

The global higher education market exceeds $2 trillion annually and is growing rapidly. By 2030, global education and training expenditure is set to reach at least $10 trillion as developing economies grow and as technology drives re-skilling and up-skilling in developed nations—the equivalent of around 6% of world GDP. Between 2020 and 2030, we expect an additional 350 million post-secondary graduates and close to 800 million more K12 graduates. Asia and Africa are where the strongest growth will naturally occur.

In 2017, the total international student market globally exceeded 5.3 million students, up from two million in 2000 (UNESCO, 2019). More than half of these were enrolled in educational programs in just six countries: United States, United Kingdom, Australia, France, Germany and the Russian Federation. Leading countries of origin include China, India, Germany, Republic of (South) Korea, Nigeria, France, Saudi Arabia and several other Central Asian countries.

The US has done a superb job in attracting talent to US schools and colleges over the last 40 years. From 1950 to 2020, the US saw a 600% increase in foreign students as a share of the total college population. International students contributed $45 billion to the U.S. economy in 2018 (source: U.S. Department of Commerce). These students typically pay higher tuition than domestic students, making many American universities today increasingly dependent on revenue streams from foreign students. Mostly from Asia, close to 162,000 international students attended colleges and universities in California alone in 2018. International student enrolment numbers declined steadily during the Trump administration, falling about 10% from 2015. The pandemic further accelerated this trend, seeing a 43% drop in new international student enrolments[6].

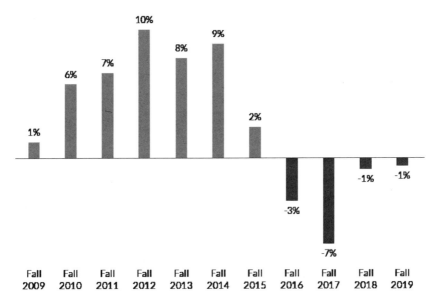

Figure 6: "Trump effect" on international student enrollment in the US (2016–2020).

Australia has an even greater dependence on foreign students than the United States—21% of all tertiary enrolments are foreign-born internationally sponsored students who collectively account for approximately $40 billion per year in cash injected into the Australian economy. According to the Australian education minister, education is Australia's "largest service-based export and supports 240,000 jobs, business opportunities and economic growth"[7].

During the 2010s, the UK was ranked as the second most popular global destination for international students after the US. In 2017, the US hosted 26% of all international university/tertiary students studying at overseas universities. The UK accounted for 12%. But the UK's market share of the international student market has been slipping more recently, with Australia, New Zealand and Canada increasing market share. Universities UK estimated that in 2014–15, international students contributed around £26 billion in gross output to the UK economy. This has become a hotly contested services component of most modern economies.

UNESCO reported in 2014 that the world needed to add 69 million teachers by 2030 to meet global education needs. But education is also ripe for structural change. For the most part, the present higher education market is still dominated by institutions that have not kept pace with globalization, most universities have not kept pace with changes in learning needs and behaviours, changes in technology, and changes in markets, and so still do not prepare what most employers would consider to be job-ready graduates.

How mass education developed

Prior to the Industrial Revolution, the classrooms we see in modern schools were rare. The education systems of Ancient Egypt, China, Greece and Rome were largely limited to training the elite, and in Europe in the Middle Ages you were most likely to be trained in monasteries or under a master artisan, rather than a scholar.

Universities first started to appear around the 11th and 12th centuries throughout Europe. The University of al-Qarawiyyin located in Fes, Morocco, is the oldest, continually operating since the 8th century, and was the first degree-awarding educational institution in the world. From the 9th to the 13th centuries, the most prized place for learning was the Grand Library of Baghdad, known as the House of Wisdom.

"Works on astrology, mathematics, agriculture, medicine, and philosophy were translated. Drawing on Persian, Indian and Greek texts the scholars accumulated a great collection of knowledge in the world... The House was an unrivalled centre for the study of humanities and for sciences, including mathematics, astronomy, medicine, chemistry, zoology and geography. Baghdad was known as the world's largest and richest city and centre for intellectual development of the time and had a population of over a million."
—Wikipedia: History of Education, "House of Wisdom"

Literacy was the primary teaching goal in most education systems, although cultural and religious concerns, along with art, music, sculpture and architecture, also were taught. Education remained largely private or church-run right up until the 18th century, when governments began to look at institutionalizing education. In 1880s France, the emergence of national secular education corresponded with the Minister of Public Instruction, who sought to break the hold of the Catholic Church and monarchism on young people.

Government-sponsored public education systems started to emerge alongside the modernization of society during the Industrial Revolution. The deployment of systemic education systems paralleled the introduction of child labour laws. Where previously schooling had been largely based on either your piety as a scholar, your academic prowess, or the wealth of your family, mass education systems needed to offer much more consistency in curriculum and methodology. Governments also saw education as a mechanism for creating orderly social and political behaviour. Indeed, many consider the modern classroom structure we see today as essentially a feeder system for the factory production lines that emerged in the early 20th century. Students attending primary and secondary/high schools were trained to be obedient drones, to take instruction from a superior, memorize a process or rule set, execute a set task, and then move on to the next subject in the production line. You were instructed to raise your hand to comment, and exams and testing reinforced process learning by rote, rather than creative thinking.

Educating for the future

While arguably largely effective for the last 150 years, it's appearing more and more likely that today's education system is not preparing our children for the next 20 years and beyond. There are several forces for change in the education sector, including the redundancy of the existing model, new technologies, new markets, shifts in employer demands, shifts in student demands, and widening gaps between the existing capabilities of teachers and institutions and market needs.

The ubiquity and democratization of education has led to a situation in which more has led to less; more programs, more students, more time spent learning; but less relevant courses, fewer students who really learn what is needed to equip them to work in the present and future worlds, and less time spent developing the skills that will be sustainable and useful in an AI-driven economy. The reason for these failings is partly financial, partly societal, and partly due to the apathy of many who control the educational futures of students.

In 2014, Elon Musk quietly pulled his five sons from Los Angeles' Mirman School for gifted children. Within the SpaceX campus in Hawthorne, California, Musk worked to establish a new type of education campus that would train his and other SpaceX employee's children for the world that is coming. Musk recruited one of Mirman's top teachers, Joshua Dahn, to run this new school, which he named Ad Astra[8].

> *"Today, more than 30 students aged 7 to 14 (including the Musk children) take part in a curriculum heavy on science, math, engineering, robotics, and artificial intelligence. Virtually everything about Ad Astra is unconventional, from its grades (there aren't any) to the secrecy of its admissions process."*
> —The three questions on the application for Elon Musk's private school, Corinne Purtill, Quartz, Nov 2018

Musk's non-profit school has done away with physical education, music and language lessons. Musk argues technology will make translation instantaneous in the next couple of years, so you're better off learning machine languages than alternate human languages. Children tackle projects like building flamethrowers and robots and launching weather balloons, along with studying nuclear politics and strategies for defeating malevolent AIs. A filing for the establishment of the non-profit with the US Internal Revenue Service (IRS) states that Ad Astra was founded to "exceed traditional school metrics on all relevant subject matter through unique project-based learning experiences". Students participate in shaping the curriculum, and annual reviews result in a very different curriculum year-on-year. A weekly assignment called Folio calls for intensive research into one particular subject. One week, it could be the cruise industry, the next, gentrification. Children that graduate from Ad Astra will have a radically different set of skills from the average middle-school graduate in the United States. But most of all, they'll be highly adaptable and engineering capable.

Beyond the handful of global universities with large endowments and the ability to adapt, many universities in the present are likely to be left behind for the following reasons:

1. World-class content will be increasingly free or low-cost access. Entities like Coursera, Udacity, Khan Academy, Codeacademy, Masterclass and others are already demonstrating how top content can be taught remotely.
2. The existing model of lectures, crowded classrooms and structured timetables will be challenged. Except in rare instances where the in-person experience or networking is a differentiator, increasingly, effectiveness of classroom teaching will be challenged. The coronavirus pandemic has accelerated this trend.
3. Academic awards will mean less and less and are likely to be seen as behind the times, out of touch, and offering a poor output in terms of graduates. We've seen companies like SpaceX, Google and Facebook waiving university requirements in recent times.
4. Community and Alumni network building remain a key selling point for universities, but very few do this well. Today this is a space where social media and online excel at.
5. New technologies like Virtual Reality (VR) could extend the classroom model cheaper and more effectively than on-campus teaching. The fastest and cheapest ways to scale education in the future will be through technology and not growing a physical campus.

Many educational institutions will need to assume the role of content aggregator, rather than aspiring to be content originators. To make the education more relevant, innovative providers could create new educational programs that leverage the degrees awarded by globally recognized institutions and twinning with professional bodies globally and with similar (already) content-rich institutions to create and bring to market learning programs that deliver job-ready graduates. In this way, students would learn relevant content and become degree qualified while at the same time attaining a relevant professional award, and plug straight into a professional community that is relevant to them and their needs.

But some question whether universities are actually important in this new world. Companies such as Apple, Google, IBM, Tesla, SpaceX, Bank of America, Hilton and others ditched college requirements over the last few years.

> *"If somebody graduated from a great university, that may be an indication that they will be capable of great things, but it's not necessarily the case. If you look at, say, people like Bill Gates or Larry Ellison, Steve Jobs, these guys didn't graduate from college, but if you had a chance to hire them, of course that would be a good idea."*
> —Elon Musk, Tesla/SpaceX

Two of the world's most gifted entrepreneurs argue that university attendance may not be a critical factor in preparing our children for employment in the world that follows. Musk has recently invited individuals to apply for jobs at SpaceX and Tesla that don't require university degrees, saying instead that he looks for "evidence of exceptional ability". As can be seen from the admission questions[9] for the AdAstra school, along with feedback from the interview process at SpaceX, Musk and his team focus heavily on problem-solving skills and thinking.

Google has gone one step further. On 14 July 2020, Google announced a series of new professional certification programs in data analysis, project management, and UX design, to be launched in collaboration with Coursera. Though Coursera typically charges a monthly $49 fee, Google is providing 100,000 needs-based scholarships to cover costs. The IT Support Specialist certification, for example, takes 3–6 months and 80% of graduates either landed a new job or benefitted from a raise. A six-month commitment costing about $300 gives graduates access to employment worth $93,000 per year—who needs university? Kent Walker, Google's Senior VP of Corporate Affairs, announced via Twitter that "in our own hiring, we will now treat these new career certificates as the equivalent of a four-year degree for related roles."

Two weeks before the Google announcement, Microsoft launched a global initiative to uplift the professional skills of 25 million people through a blog post on 30 June 2020.

Jack Ma has also said that the current education system is woefully ill prepared for the skills required to thrive over the next 20–50 years. Ma suggests that there should be much greater investment in children at younger ages, when kids are building skills and values, and much less in universities, when values are already hardcoded. Ma claims that Alibaba and Ant Financial regularly find themselves retraining university graduates, and he's been quoted as suggesting that a university degree was nothing more than a "receipt for the tuition paid".

In an interview we conducted with Jack Ma for our book, Jack emphasized the difference between education modality of the 19th century and the 21st century will be in the ability to differentiate humans from machine capability. As such, Ma argues, as does Musk, that current education systems are reinforcing learning knowledge by rote, rather than the wisdom of application of that knowledge.

"The future is not just about knowledge, but about the wisdom to apply that knowledge. The key to competitive differentiation between AI and humans lies in our mission and values. Human knowledge has most certainly exploded sharply, but our wisdom has not improved for thousands of years. Knowledge can be transferred to an algorithm,

but wisdom is based on aggregate experiences.
You can be clever based on learning, but you only get
smarter when you incorporate experiences."
—Jack Ma

Evidence of this can already be seen in Europe today, where Finland has topped EU rankings for their comprehensive school system for the last 16 years unchallenged. Children start schooling very late in Finland compared with other school systems, often not attending primary school until the age of 7. In preschool children aren't taught any math, reading or writing; the emphasis is on creative play and developing good social skills, like making friends and respecting others. Teachers emphasize this is learning through play—this in turn builds a lifelong appreciation of learning rather than an apprehension around meeting grades. Students do not sit exams prior to university at all.

One aspect of education that will be critical is competing against Artificial Intelligence. We talk often about STEM competencies today: Science, Technology, Engineering and Math. However, technology won't be a separate discipline in the near term, it will be embedded in every program, every course. Living and working with robots will be a skill that differentiates you in the developed world, but will also be critical to new infrastructure deployment in developing economies in places like Africa. Energy systems will be smart. Cities will be smart. Healthcare, transport, farming and supply chain will all be underpinned by large-scale automated smart systems. In this way, education will have to be smart, too. Soft skills like creativity, EQ and LQ (Love Quotient), as Jack Ma puts it, will differentiate us from machines; but everyone will have to work with AI in some capacity. Hence, students will be exposed to technology from the earliest age, and won't choose technology as a separate discipline, and coding will be like math today—taught at all levels of basic schooling.

Ma argues that love and human fellow feeling and compassion will ultimately differentiate us from the logic and intelligence of AI.

"Yes, our AIs will have phenomenal intelligence built into their neural circuitry, but only humans have great hearts and the capability to love. Machines have great precision and accuracy, but people have colour, nuance, and temperature. The age of AI will usher in the age of human togetherness. Only in this way will humans not be eliminated as the world transforms augmented by technology.
Machines might be able to replace a babysitter, but clearly not a mothers' love for her child. Machines are already replacing nurses who dispense medicine and surgeons who operate on patients, but they can't emulate compassion and concern for the injured, sick and infirm. In the era of AI, if there is no love or humanity injected into business, the stronger the reliance on technology, the more clinical, more injurious and inequitable it will be."
—Jack Ma

The real promise of the future of education though, is dramatically lower costs and increased teacher effectiveness. As teachers get plugged into technology-based distribution systems like VR, we can expect that teachers will be able to increase their earnings as they use assisted teaching systems to reach greater class sizes. Some face-to-face collaboration, social skills and team-building requirements will still require students to gather physically, but this could be based on roster systems or embedded community mentoring and facilitation. By the second half of the century, we imagine that preschool will be classroom based, but that much of the middle years of schooling from ages 9–16 will be handled using a combination of in-person and virtual teaching systems. This will make state-based schooling not only far more effective from a cost perspective, but also far more inclusive.

The end of homelessness

Inequality in cities like San Francisco has produced incredible pressure on lower- and middle-income families in respect to housing affordability. In the most severe cases, this has resulted in an epidemic of homelessness that plagues the Mission district and other areas of San Francisco, or Skid Row in Los Angeles. But it's not just in extreme cases like the Mission district or Skid Row where homelessness is an issue.

Based on aggregated national reports compiled by Habitat for Humanity, the UN and others, it is estimated that more than 150 million people globally suffer from long-term homelessness, while 1.6 billion lack adequate housing—that's about 20% of the world's current population.

Manila, the capital of the Philippines, has the highest concentration of homeless people in the world, with an estimated 3.1 million[10] people living unhoused—that includes 1.2 million children. Recent evidence suggests that more than two million are homeless today in the US due to the economic impact of the coronavirus pandemic. A third of this total are displaced families. More significantly, in the US during the coronavirus pandemic, tens of millions of households were unable to meet their rental commitments. The Aspen Institute estimates that 30–40 million households face potential eviction still[11].

Homelessness is closely correlated with housing affordability, which is why cities like San Francisco and Los Angeles have high homeless populations. In cities where people spend more than a third of their income on rent, recent research in the US shows that homelessness increases in severity[12].

In the United States, the annual cost of policing, feeding and providing emergency physical and mental health support to a homeless individual is reported to cost $35,578 per person per year. Recent research in Orange County[13] and various cities have shown dramatic reductions in cost by providing more structured environments for those unable to house themselves, with far greater propensity for reengaging such individuals in gainful employment. In fact, the US Department of Housing and Urban Development reports that the average costs to government (in respect to homelessness) would be reduced by 49.5% if the homeless were placed in supportive housing. A study conducted by University of Melbourne, Australia, showed that just 19% of Australian homeless are employed, and they were statistically 30% more likely to leave their job than a typical

Aussie. In King County, Seattle, 50.2% of homeless surveyed in 2018 had been unemployed for the preceding 12 months, while 64% said their current episode of homelessness had lasted a year.

Companies like Icon, Sunconomy, Baby Steps, APIS Cor, XTreeE and CyBe Constructions have developed technologies for 3D printing basic one- to three-bedroom homes in 10–48 hours. Icon, for example, uses similar technology to the 3D printer you might have at home, using pumps that extrude Lavacrete II (A Portland cement-based mix with proprietary ingredients). It is said to have a compressive strength of 6,000 psi, putting it well above most standard building materials today. In fact, 3D-printed homes are cheaper and faster to build, are eco-friendly, are generally stronger than current construction techniques and fare better in earthquake and fire tests.

The startup AI SuperFactory recently won NASA's 3D-Printed Habitat Challenge when they 3D printed a primitive demonstrator of a Martian habitat building in less than 30 hours. Their concept of a Mars Hab called Marsha was designed to utilize in-situ resources from the Martian environment, enabling robots sent to Mars to cook up a 3D-printed biopolymer basalt composite that when cured is 50% stronger and more durable than concrete. This raw material is also recyclable.

The Additive Manufacturing Integrated Energy (AMIE) demonstrator is a project of the US Department of Energy's Oak Ridge National Laboratory. The aim of this 3D-printed home was to attach an energy-independent building design to a hybrid electric vehicle, creating an integrated energy system. The integrated PV solar cells provide energy for the structure, charging the vehicle during the day, and the vehicle acts as a store device to extend energy hours for the home in the evening.

Micro-unit apartments are also increasingly popular for urban planning as population densities in cities increase, along with rental costs. Cities like Hong Kong, Sydney, New York City and even Austin, Texas, are trialing micro-apartments that range from 10–16 square feet (1–1.5 sq metres, or smaller than a prison cell), through to 70–250 square feet (6.5–23 sq metres). The cage homes of Hong Kong are just 4 square feet (0.37 sq metres) on average. However, the micro-apartments being developed increasingly use a ton of tech to make the spaces more livable.

Figure 7: 3D-printed homes on Earth and Mars. (Image credits: Icon/AI SpaceFactory)

The point is, that we have dozens of companies around the globe now creating homes that cost under $10,000 to build and deploy (many under half that cost). Based on the financial impact of homelessness to society, there's a very strong argument that zero homeless should be the goal for every developed nation in the world today. We also have to plan for higher urban populations and better use of city spaces to accommodate these growing populations.

As these construction costs continue to decline, there is a point at which it simply becomes uneconomical to continue to allow homeless individuals lack of access to basic housing. Regardless of where you sit on the political spectrum, it is simply cheaper to house the millions of homeless than leave them on the street.

Financial inclusion via digital inclusion

Today, it is estimated that around two billion people worldwide have no access to the types of basic financial services delivered by regulated financial institutions and banks. Historically that figure has actually been even higher. Until just a decade ago, around half of the world were unbanked and considered by most financial institutions to be unbankable because they didn't meet the classic threshold of profitability.

In 2005 if you lived in Kenya there was a 70% chance you didn't have a bank account, nor could you store money safely, and most likely your savings were non-existent. Today, if you're an adult living in Kenya there's a 98% likelihood that you have used a mobile money account (stored in your phone SIM), and that you can transfer money instantly to any other adult in Kenya. Data shows that Kenyans trust their phone more than they trust cash in terms of safety and utility, with people sewing sim cards into their clothes or hiding them in their shoes so they can more safely carry their money with them. This is all possible because of a mobile money service called M-Pesa, created by the telecommunications operator Safaricom. Currently at least 40% of Kenya's GDP runs across the rails of M-Pesa[14].

> *"We're currently sitting at about twenty-two million customers out of a total mobile customer base of about twenty-six million. Now, if you take the population of Kenya as being forty-five million, half of whom are adults, you can see we're capturing pretty much every adult in the country. We are transmitting the equivalent of forty percent of the country's GDP through the system and at peak we're doing about six hundred transactions per second, which is faster and more voluminous than any other banking system."*
> **—Bob Collymore, CEO of Safaricom/M-Pesa[15]**

When it comes to financial inclusion, Kenya has done more to improve the lot of its populace in the last 10 years than the US has in the last 50 years. Indeed, Kenya today has higher financial inclusion than the United States—a mind-blowing and clearly inconvenient statistic. In the US the Federal Reserve reports that approximately 20% of US households are unbanked or underbanked. Yet, the United States has one of the highest densities of bank branches in the world. How do you get one of the highest density of bank branches and still have one fifth of households underbanked? The answer is identity.

One of the chief causes of financial exclusion today isn't access to banking, but access to the identity documents that are required for opening a bank account or voting. Since 9/11, documentary requirements in the

United States to open a bank account have become stricter, in line with the Patriot Act and the Customer Identification Program (CIP) enshrined in US banking law and regulations. However, more than half of the US population doesn't have a passport (only 42% as of 2018[16]), and only 76% of the population has a driver's license. Even if you can get to a bank branch you still might not be able to open a bank account.

In India, up until 2014 less than 30% of the population had a bank account. The Reserve Bank of India had tried increasing branch access—in fact, they put in place regulations that meant growing banks in India that wanted to deploy new branches had to put a quarter of their new branches in rural areas not currently served by a bank. This policy was in place for almost a decade, but hardly moved the needle on financial inclusion, before Nandan Nilekani (the co-founder of Infosys) explained to President Modi that the problem wasn't just access to bank branches, but access to an acceptable form of identity that you could use to open a bank account[17].

This is why India's initiative to deploy the Aadhaar card was so critical to enable financial inclusion—it changed the game. As of 2017, more than 1.17 billion have been enrolled in the Aadhaar card program. That's 88% of the Indian population. The effect of identity reform in India is that the number of those included in the financial system has skyrocketed. The segment of the population most excluded in the old banking system—lower income households and women—have seen 100% year-on-year growth since the Aadhaar card initiative was launched. As of 2015, more than 358 million Indian women (61%) now have bank accounts, up from 281 million (48%) in 2014. This is the biggest single jump for 'banked' women among eight South Asian and African countries. You can either lower identity requirements or create new identity structures to support inclusion, but you can't create identity verification requirements that require drivers' licences or passports for a population that doesn't drive and doesn't travel. Bank branches are useless in these scenarios, because even if you get someone that is financially excluded into a branch, they still won't qualify for a bank account. That model is a recipe for financial exclusion, as the 25% of US households that are underbanked already know.

Research by Standard Bank and Accenture back in 2016 concluded that of the approximately one billion unbanked in sub-Saharan Africa, 70% of those individuals would need to spend an entire month's salary just to physically get to a bank branch. This statistic clearly indicates a significant structural problem in solving financial inclusion on the continent if left to traditional banks.

In Kenya, where approximately 48.76% of GDP flows through M-Pesa[18], Kenyans are reported to be saving up to 26% more today than when they only used cash. As a result, 60% of Kenyan's trust M-Pesa more than cash today. Crime is down, savings are up, but the more interesting effects are in response to poverty, credit access and employment. Access to mobile money lifted 2% of Kenyan households (194,000 families) out of extreme poverty, 185,000 women out of subsistence farming into business, and increased access to basic credit facilities for starting a business or dealing with emergencies[19], as examples.

It's clear that financial inclusion should be a basic goal of the global economic system, but banks are deterred from this goal due to lack of profitability from serving lower-middle income segments and lack of digital inclusion. If commerce is going to continue, equality alone dictates that access to basic financial services is a necessity for all. However, the banking system that was created by the Medici's of Italy in the 14th century hasn't solved basic access to financial services in more than 500 years. While bankers told us that bank branches was the solution to access for financial services, countries with the highest density of bank branches like the United States, Spain and France continue to have lower financial inclusion than that of Kenya today! Kenya and India grew from less than a third of the adult population included in the classical banking system, to somewhere north of 90% in just a handful of years—but none of that was due to banks or bank branches. It came down to two simple changes: the creation of national identity schemes and access to basic mobile phone technology. These two mechanisms are responsible for the greatest ever financial mobilization the world has ever seen.

The next stage of this revolution will be the creation of commerce that sits exclusively on top of the mobile internet. Digital inclusion should now be one the primary goals of governments the world over, recognizing that financial inclusion, commerce and growth are closely aligned to the evolution of the smart phone and the internet. Digital inclusion looks like it will become a basic human right along with access to electricity, sanitation, fresh water, education and basic healthcare.

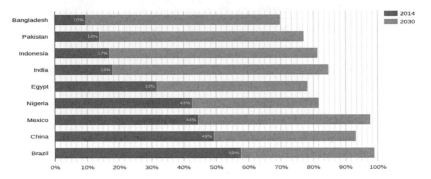

Figure 8: Internet penetration by 2030. (Source: UN/Wikimedia)

By 2030, it is anticipated that more than 95% of the world's population will have access to the internet via mobile. Smart phones have increasingly become cheaper to manufacture and deploy. Today, brand-new basic smartphones can be found on the streets of India, South Africa and Nigeria for under US$40. By 2030 it is expected that such devices will be available essentially for free with basic subscription services for access to the internet—by 2025 Nigeria will reach 65% smartphone penetration. It's expected that tech giants like Facebook, Google, Tencent, Alibaba and Amazon may give smartphones to individuals who subscribe to basic services through their infrastructure. By 2050 access to basic internet infrastructure will be ubiquitous across the planet, meaning everyone will participate in the services available in the digital economy.

Universal healthcare through rethinking data and disease

Across the developed world, basic healthcare has been improving for most citizens over the last 100 years. Infant mortality has plummeted. Life expectancy has more than doubled since the 1850s. Diseases we once thought incurable, such as smallpox, have almost been eradicated, while other diseases like Ebola now have vaccines that prevent the death of tens of thousands. Access to a doctor or hospital is as simple as getting in a car and driving into town. For most, that is.

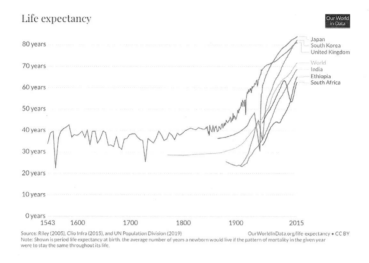

Figure 9: Improvements in life expectancy have mostly come in the last 100 years.

(Source: Riley, Clio, UN)

The United States has resisted this trend for improved access and affordability to healthcare, being the single developed nation in the world not to have a universal healthcare system and with the most expensive healthcare costs of any nation today. Over 65% of all bankruptcies in the United States are reportedly related in some way to medical costs[20]. There is a rational argument to be made that no one should have to die because they can't afford healthcare in the richest, most prosperous economy in world history. Yet it is estimated that 45,000 people die each year[21] in the United States from preventable conditions simply because they don't have access to basic health insurance or affordable medicines. The US is not the only nation considering the ongoing feasibility of universal healthcare. Countries like Australia, UK and others are finding escalating healthcare costs to be a battle that at some stage might be lost.

The United Nations considers healthcare a basic human right. More than half of the world's countries have pledged to protect their citizens' right to healthcare, either through human rights agreements or national laws. Pew research in 2018 showed that more than 60% of Americans think it is the government's obligation to guarantee healthcare support. Unlike the rest of the G20, the US does not have a national healthcare system, it has a national health insurance scheme—which places the reliance on employers and individuals to care for an individual's health. Not the government.

The political argument often voiced in the US is that if an individual can't afford healthcare they just aren't working hard enough. But at the heart of this argument is a system which is simply inaccessible to a large portion of the population due to high private care costs. There are people working three jobs, working 16–18 hours a day, who still can't afford basic healthcare—and you can't argue they aren't working hard enough. In 2020, the average cost for health insurance in the US was $456 for an individual and $1,152 for a family per month. On the minimum wage of $7.50/hr, that would represent 91.5% of an individual's monthly salary. That's even before we factor in that someone on minimum wage can't even afford to rent a one-bedroom apartment anywhere in the US today.

For many in the US, the employer is the sole arbiter of whether an individual can get access to healthcare. Many large employers like WalMart hire a high percentage of part-time workers so they can avoid healthcare costs. To be eligible for healthcare at WalMart you must have been at the company for at least one year, and you must work 36 hours per week or more[22]. That excludes about half of WalMart's workforce today[23]. For most developed nations around the world, 30–35 hours per week would be considered full-time employment for the purpose of benefits.

In the 1980s, costs for healthcare in the US started to balloon. Before the 1980s payments by Medicare and insurers were tied to procedure costs—if it cost a hospital $5,000 to do an appendectomy, then that's what the hospital was paid (with some margin for profit). But in the early 1980s insurers and the Reagan government began to shift financial risk to health providers. Medicare began to pay hospitals a fixed price per visit. These squeezed margins dramatically, so hospitals and doctors began to search for revenue at every turn.

In the 1990s healthcare reform was a subject of much discussion in the

US as the cost of providing healthcare continued to accelerate as a budget line item for the government. The push to privatise the system was initially argued to be successful, but since then costs have continued to balloon, leading to large-scale exclusion of citizens from access to basic healthcare coverage.

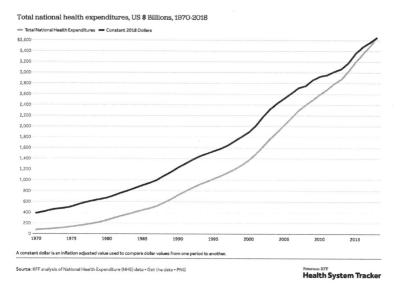

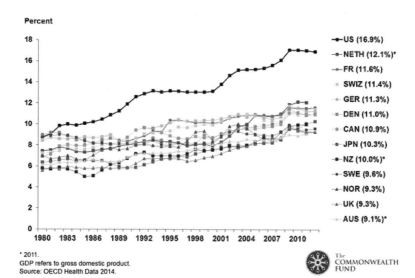

Figure 10: Total health expenditures, last 50 years
(Source: NHE Data) vs OECD countries (Source: OECD).

One of the key reason's healthcare costs grew to approximately 17% of GDP in the United States wasn't necessarily a better quality of healthcare, but the increasing quantity of care per patient. In the US the emergence of for-profit health care resulted in a high probability of unwarranted procedures, and new services and fees. The US, for example, has the highest rate of invasive (and risky) cardiac procedures in the world—45% more than the next country. Yet, all these additional diagnostic procedures have not bought Americans better heart health—in fact, the US has declined on many metrics when compared with other modern economies:

- Access and quality to healthcare in the US was lower than the Netherlands, Australia, Sweden, Japan, Germany, France and the UK for the same per capita spend[24]
- Chronic disease burden is much higher in the US compared with other leading economies (almost double that of Japan), as measured by age-standardized Disability Adjusted Life Years (DALY) rate per 100,000 population—and obesity is a significant cause of higher chronic disease
- Hospital admissions (as opposed to day care) for preventable diseases are 37% higher in the US[25] for four major disease categories (congestive heart failure, asthma, hypertension and diabetes)
- The US has higher rates of medical, medication and lab errors than other leading economies, affecting 19% of US adults, versus just 7% in countries like Germany and France[26]
- The mortality rate for respiratory disease is higher in the US than comparable countries
- Adults in most leading economies have quicker access to care than in the US
- Use of the emergency department in hospitals (for conditions treatable by regular GP) in the US is three times higher than France, UK, Australia, Netherlands and Germany
- Administrative process in the US, at 35% of the total annual healthcare cost, is double that of the OECD average

In most of the developed world, healthcare is seen as a basic function of government, and something that taxes should most definitely pay for, along with education, roads, airports, sanitation, energy and such. In economic terms these are identified as public goods—the basic elements of society that the government provides its citizens. In the United States, and more recently in populist politics, right-wing conservatives have attempted to shift healthcare out of the basic or public goods classification, into the realm of personal responsibility for individuals.

The broad argument is that the costs of healthcare is one of the single biggest social impact costs for society, and that it is simply too expensive a burden even for more efficient forms of government to bear. The argument being that you can't have lower taxation and smaller government footprints and still maintain services like healthcare for the general populace, so in return for more efficient government (read: lower taxes) we need to accept lower success rates on inclusive healthcare. This argument makes the assumption that the cost of healthcare will continue to rise. Thankfully, we now understand that this is unlikely to be the case beyond 2030, perhaps even earlier.

The key issues for escalating healthcare costs globally include:

1. **Misdiagnosis**: Misdiagnosis costs the US economy an estimated $750 billion a year (around 3.5% of GDP), and is the cause of 40–80,000 deaths per year.
2. **Late diagnosis**: Recent studies have shown that early diagnosis has the potential to dramatically reduce overall treatment costs.
3. **Drug development costs**: While necessary, regulation has led to dramatic costs in developing drugs and making them available to the public with FDA approval in the US skyrocketing to $2.6 billion per new drug, with only 12% of them making it through clinical trials.
4. **Administrative costs and closed loop systems**: An Optum study estimated that $200 billion in administrative waste is generated annually that US healthcare payers and providers are unable to reduce on their own, but that AI could eliminate.

5. **Aging population and longevity improvements**: By the time you reach 65 years old, average healthcare costs are presently $11.3K per person, per year.

6. **Obesity and poor diet**: The Milken Institute estimates the total cost of chronic diseases in the US due to obesity and people being overweight was $1.72 trillion (equivalent to 9.3% of the U.S. gross domestic product in 2019).

As discussed in *Augmented*[27], Health Tech is exploding right now, and it promises a shift in the way we think about healthcare and the costs associated with it. However, it requires a fundamental reorganization in respect to the medical sector, government regulations and big pharma in order to capture these cost benefits.

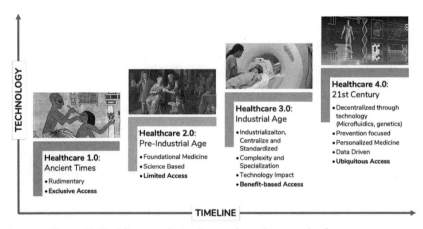

Figure 11: Healthcare evolution has accelerated as a result of computing.

Much better healthcare at much lower costs

The first sea change in healthcare is coming through the field of **gene-therapy**. Previously, we diagnosed diseases by looking at symptoms. Often those symptoms could be masked, making diseases difficult to diagnose accurately, or those that did present could apply to any range of conditions. However, when we eliminate viral and bacterial infections, hereditary diseases encoded in the genome are increasingly becoming much easier to diagnose. If you have certain genes present, when married with symptoms, diagnosis becomes much more precise. In November 2019, Matt Hancock, the United Kingdom's health secretary at the time, announced a plan to sequence the genome of every baby born in a National Health Service hospital, beginning with a pilot of 20,000 children. Hancock called his plan a "genomic revolution," promising that whole genome sequencing and genomics would play a huge part in ensuring that every child receives "predictive, preventive, personalized health care".

Countries like China, Japan and Korea have already started programs to sequence the genome of every child born to get these baseline data sets for future healthcare of citizens. The cost of gene sequencing has come down 10,000-fold in just the last 10 years and continues to decline.

FIGURE: FALLING COSTS OF HUMAN GENOME SEQUENCING (2001-2017)[14]
Source: NIH, National Human Genome Research Institute, Apr 25, 2018

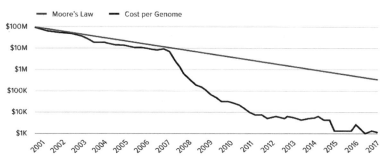

Figure 12: Dramatically falling costs of gene sequencing.
(Source: NIH, National Human Genome Research Institute Apr 2018)

Gene sequencing will enable dramatic improvements in diagnosis as we will be able to align symptoms with inherited genetic traits, and eventually this will lead to gene-therapy which eliminates well known diseases from our genome.

The second sea change is in **diagnosis, data collection and modeling technology**. Computer imaging, microfluidics, laser interferometry and modeling techniques combined with the use of Artificial Intelligence is producing a revolution in diagnosis capability. While the controversial health-tech startup Theranos was ultimately unsuccessful at commercializing microfluidic diagnostics, companies like HP are now working on handheld labs that can instantly diagnose a range of blood-borne conditions in real-time. We're beginning to understand the role that bacteria, gut microbiome, plaque and toxins have in a range of conditions, largely as a result of better data and computer modeling. Rapid diagnostic improvements have come through Moore's law—increased computing power has led to breakthrough after breakthrough in medical research.

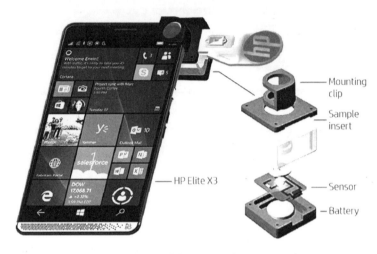

Figure 13: HP Diagnostic labs integrated into smartphones.
(Image Credit: HP Healthcare)

The third area where medicine is going through a fundamental change is in **treatment approaches**. As we combine gene sequencing, regular and inexpensive blood work, sensors on and inside our body combined with artificial intelligence to mesh this data together, medicine will shift to largely preventative approaches. We won't need to wait for symptoms to appear—in fact, the equivalent of our smartphone in a decade's time will likely know when we're sick well before we do. When it comes to treatment of emerging health issues, we will be using both gene-therapy to turn on and off offending proteins and gene switches to remove serious conditions from our DNA, using the body's own system to produce antibodies and the required proteins to prevent disease development. When intervention is required, such as in the case of cancer or a serious viral infection, we'll be able to produce personalized medicines based on an individual's genome (this is known as pharmacogenomics) and the genome of the cancer or virus itself. This era of personalized medicine will obviously be very different from the mass-produced, patented drugs that we find on pharmacy shelves today.

While many technologies today are extremely expensive and have added to the cost of healthcare over the last 20–30 years, Artificial Intelligence is the key to rapidly declining diagnosis and treatment costs as we get much better at understanding systemic responses and the most effective treatment regimen for the individual. Accenture research demonstrated the potential of AI in healthcare to realize $150 billion in annual savings by 2026, just in the United States alone.

We're going to use a range of technologies to dramatically reduce treatment costs and improve diagnosis, including:

- **Virtual and Augmented Reality Technology**—from surgeons that wear A/R glasses to improve the accuracy of spinal, cardiac and cancer surgeries, through to VR being used as alternative treatments for PSTD, stress and mental health conditions.
- **Robotics**—from robotic surgery and assisted surgery, through to tele-operated robotic nursing assistant units that travel wards dispensing medicine and checking on patients.
- **Wearable and Ingestible Sensors**—from wearable heart rate sensors in watches that have already detected abnormal heart conditions and saved lives, through to ingestible sensors that could track blood pressure, blood sugar levels and even regulate insulin, the marriage of real-time health data with AI diagnosis will be massive for improving the cost of treatment and early detection.
- **Tele-health and Remote Health Management**—diverting patients from emergency departments through tele-health can save more than $1,500 per visit, according to research from *The American Journal of Emergency Medicine*[28]. Using personal health apps in our smart ecosystem to determine whether we use telehealth consultation or go to an emergency room, will improve system efficiency massively.

- **AI Diagnostics**—AI diagnosis in various fields is already performing at the same level as humans, and it won't be long before it's performing at significantly higher rates of accuracy. Lancet Digital Health analysis of 25 diagnostic AI studies found that machine learning correctly diagnosed diseases at the same or better rate than human healthcare professionals. The specificity for deep learning algorithms was 93%, compared with humans at 91%.

- **Medical Tricorders**—from Google's Verily moonshot spin-off to the DxtER Medical tricorder device, we're seeing improvements in handheld computing power and sensors get us very close to a working diagnostics device. This would allow a handheld unit to diagnose hundreds of health conditions based on genome, family medical history and real-time sensor readings.

- **3D Bioprinting**—by the end of this decade 3D bioprinting will allow us to replace organs, augment orthopedic reconstruction, and potentially augment organ function. We have already successfully produced 3D-printed bladders, esophagus and kidneys, and used 3D printing for facial reconstruction surgery. In 2020, engineers produced the first fully 3D-printed heart models for surgeons to practice on, but by the later part of this decade we should be able to address the fine vasculature required for complex organs like the heart and liver.

- **Nanotechnology**—from 2030 to 2050, we will be experimenting and perfecting the use of medical nanobots, or tiny, microscopic robots that can treat you from inside. Experimental nanobots can already deliver highly targeted drugs to a very specific area, say, delivering cancer drugs to a cancerous growth or tumour. By 2050, nanobots may be able to repair cellular damage, repair broken bones, torn muscles, damaged blood vessels and so forth in real-time.

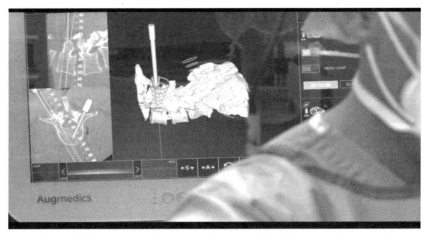

Figure 14: Surgeons are already using Augmented Reality in spinal, heart and cancer surgery. (Image Credit: Augmedics xvision™ Spine System)

The biggest improvements in overall healthcare cost effectiveness will come through reducing treatment variability and healthcare practitioner hours. In some procedures in the US, 30% of the overall cost is carried by just 1% of patients[29]. Artificial Intelligence in diagnostics is already showing that deep learning reduces error rates significantly and performs better than 95% of doctors in respect to diagnostics on cancer, cardiovascular imaging, and even for something as basic as echocardiogram (ECG) analysis. Shared medical data will also reduce the incidence of unnecessary procedures and false positives, as medical history will be much more precise. When combined with historical genetics, the ability to narrow down treatment regimens based on germline conditions versus viral or bacterial infections will be far superior than the methods we utilize today. It will be more precise by magnitudes.

"Waste in medical care comes in many forms.
One clear cause is misallocated treatments:
spending on care that is not clinically valuable or
not spending on preventive services."
—David Cutler, Professor of Applied Economics at Harvard University

Thus, by the 2030s when we're looking at the potential for dramatically optimized national healthcare, we'll be data oriented and powered by AI in our ability to create national systems of health management. People will live longer, healthier lives and we'll shift to preventative health maintenance as a primary modus operandi, versus symptom management and mass-market drug production. Health insurance will be replaced by health management services on a subscription basis, with differing levels of care dictated by data sharing around your genome and medical data.

In current models, healthcare costs are expected to increase annually at 0.8% above GDP increases. During the period 2000–2019, before COVID-19, annual healthcare expenditure increase in the US was 5.6%. Reducing government spending isn't the answer, as National Institutes of Health (NIH) data shows that just 3.86% of annual spending is government administrative costs. But there is plenty of potential for lowering the costs through technology-based reform.

By the late 2030s and early 2040s, healthcare per citizen for countries that centralize national health data schemes will be potentially half of what it costs today. This will come through the following mechanisms:

1. Gene-therapy for chronic disease (18–30% reduction in total annual treatment costs[30])
2. Automation of administrative workload (10–15% saving[31])
3. AI application to repetitive tasks/diagnosis (20–60% saving[32])
4. Better targeted treatment and personalized medicines (20–30% increase in efficacy)
5. Preventative/early treatment improvements (30–40% systemic cost reduction[33])
6. Reduction of obesity-related issues ($450–1.7 trillion (2018) in potential savings)
7. Better chronic care management (20–30% reduction in Medicare costs alone)

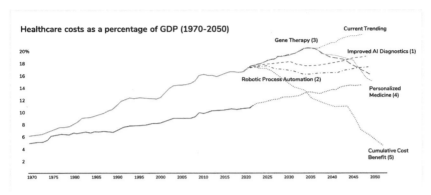

Figure 15: Potential reduction in national healthcare costs using tech. Source: Author's own based on origina NIH and OECD Data applying trending analysis and potential cost benefit of AI, Gene Therapy and associated technological advances

Notes: (1) Improvements in diagnostics with the use of AI is likely to reduce US healthcare spending by at least 20% or 3.5% of GDP by 2050 at a minimum. (2) Robotic Process Automation will eliminate at least 80% of administrative costs resulting in as high a reduction as 40% by 2050, although we've discounted that slightly due to legislative lag. (3) Gene therapy will reduce or eliminate the treatment costs of many genetic diseases we experience today such as Alzheimers, Parkinsons, multiple cancers, Inherited Cardiac disease, etc. Total cost reduction potential is north of 40% but it won't start to kick in until the early 2030s. (4) Personalized medicine will be a digital health service layer based on preventative treatments combined with real-time diagnosis. Drugs and medicines will be individualized based on your genome, bloodwork, gut biome analysis, and the genes of viruses and bacteria realizing much greater treatment efficacy.

The most advanced smart economies will no longer see rising healthcare costs with tech-based reform of healthcare. In fact, once the investment in these core technologies and systems is in place, national data pools combined with privatized health management services could cut the cost of healthcare at a national level by 30–50% annually, all the while massively increasing the efficacy of outcomes.

This will enable us to look at the biggest disease challenge of the 21st century, namely ageing. By 2040 it is estimated that we will reach "longevity escape velocity", where we will be able to first neutralize ageing, and then reverse age-related health effects. In 2015 a team of international geneticists and scientists wrote a paper titled "It is time to classify biological aging as a disease" (Bulterijs et al, 2015). They argued that historically, ageing has been seen as a natural process, but scientific research over the last two decades has shown that we can slow or reverse elements of that natural process; therefore, clinical ageing will eventually be treated like any other preventable or manageable condition.

In confirmation of this, in 2018 the WHO added to the International Classification of Diseases database a specific extension-code identifying "ageing-related" diseases—defined as those "caused by pathological processes which persistently lead to the loss of organism's adaptation and progress in older ages".

Aubrey de Grey
@aubreydegrey   ...

I now think there is a 50% chance that we will reach longevity escape velocity by 2036. After that point (the "Methuselarity"), those who regularly receive the latest rejuvenation therapies will never suffer from age-related ill-health at any age.

3:29 AM · Mar 15, 2021 · Twitter Web App

Figure 16: Aubrey de Gray, one of the world's foremost longevity scientists, predicting 2036 as the year we start to de-age. (Credit: Twitter @aubreydegrey)

We could write an entire book on the impact of longevity treatments and the end of ageing, but we'll leave this for now. We will explore longevity as an element of society in the later chapters of the book.

For certain services, life extension therapy for example, we still expect inequality to be influential, with the rich getting access to these expensive treatments first or having home-based diagnosis capabilities and access to drone-based delivery of personalized medicine, bio-printed organ replacement options, etc. However, our ability to manage the health of our citizens at much lower costs is the primary outcome of these systemic changes and the evidence seems clear. Healthcare will be radically cheaper once we apply technology reform across the sector.

From 1975 the national cost of healthcare as a percentage of GDP has risen across OECD nations. By 2030 it will start to flatten. By 2040 it will most certainly be in decline.

It's not socialism, it's a business case

Ultimately, this is about simple economics.

While the 1980–2010 period in developed economies created booming private healthcare, private education and financial systems that were extraordinarily successful for the wealthy, the obvious failures across the social spectrum means we recognize today that the free market has tended to create increasingly polarizing outcomes, depending on your economic standing. This is counter-intuitive to both the central purpose of the economy itself and the philosophical drivers of humanity as a whole.

The greatest social issues of our time, namely education, healthcare, financial inclusion and homelessness, have technical and strategic solutions that are likely to dramatically decrease the cost of providing those basic needs to our citizens over the next two decades. There will come a point in the second half of this decade where politicians we used to label as socialists will be able to competently argue that nation states can eliminate social probems through leveraging large-scale systems based on massive data sets and AI. It is inevitable that once most of the population believes that these solutions are within our grasp without additional tax burdens, that resistance to more socialized policies will decrease.

Universal healthcare and housing the homeless under technosocialist principles will be far more cost effective than leaving it to a disincentivized, ineffective private market. Education will be far more accessible and higher quality education will be far more available by 2035. Ultimately, the cost of providing these basic goods will be a much lower per capita cost (GDP) than today thanks to technology.

As automation impacts employment, the likelihood of medical bankruptcies, reduced access to private healthcare and distressed access to basic housing, education and food, will be enormous. The mitigation of social revolt will require national commitments to providing these services across the population as a basic element of economic output. Given a choice of investing in technology reform to provide basic goods like healthcare, housing and education at much lower costs, versus simply leaving it to free markets to manage, governments may find themselves increasingly rejected by the populace as unable to solve basic problems. The prospect that citizens will simply accept an environment where policies are not developed to deal with inequality and exclusion is increasingly unlikely. The level of transparency afforded us by the internet has created a much more informed and involved citizenry. There is a point where revolution is not only likely, but probable, if these basic services aren't baked into the economy.

Endnotes

1   Grand Old Party, otherwise known as the Republican Party.

2   Source: Brookings Institute, "The Climate Crisis, migration and refugees", 25 July 2019 (https://www.brookings.edu/research/the-climate-crisis-migration-and-refugees/).

3   See Centre on Budget and Policy Priorities – "Immigrants Contribute Greatly to U.S. Economy, Despite Administration's "Public Charge" Rule Rationale", 15 August 2019.

4   Source: OECD Report, "How Immigrants Contribute to Developing Countries' Economies", OECD/ILG 2018 (Chapter 5: Immigration and Economic Growth).

5   Source: Union of Concerned Scientists, International Energy Agency combined with Earth Systems Scient Data 11, 1783-1838, 2019, 2020: See https://www.ucsusa.org/resources/each-countrys-share-co2-emissions.

6   Source: Institute of International Education (IIE) Report, "Fall International Enrollments Snapshot Reports", November 2020.

7   Source: Minister of Education, Media Release, 22 November 2019: https://ministers.dese.gov.au/tehan/international-education-makes-significant-economic-contribution.

8   Latin for "To the stars".

9   See Quartz.com, "The three questions on the application for Elon Musk's private school", Corinne Purtill, 30 November 2018: https://qz.com/1480109/the-three-questions-on-the-application-for-elon-musks-private-school/.

10 Source: Reuters.

11 Source: The Aspen Institute, "The COVID-19 Eviction Crisis: an Estimated 30-40 Million People in America Are at Risk", Benfer et al, August 2020.

12 Source: Homelessness Rises Faster Where Rent Exceeds a Third of Income, An Analysis by Zillow Research Fellow Chris Glynn of the University of New Hampshire, Thomas Byrne of Boston University and Dennis Culhane of the University of Pennsylvania, December 2018.

13 See "Homelessness in Orange County – the costs to our community", June 2017, University of California, Irvine; United Way and Jamboree.

14 Source: The Economist, "A new East Africa campaign", 9 July 2015.

15 Source: BreakingBanks Fintech Radio Show.

16 Source: Census Data and VOA/Forbes (see https://www.forbes.com/sites/niallmccarthy/2018/01/11/the-share-of-americans-holding-a-passport-has-increased-dramatically-in-recent-years-infographic/#a7050043c167).

17 "Nandan Nilekani impresses Narendra Modi & Arun Jaitley, gets Aadhaar a lifeline". The Economic Times, 24 July 2014.

18 Source: Payments Cards & Mobile, 25 January 2019: https://www.paymentscardsandmobile.com/mobile-money-transactions-half-of-kenyas-gdp/.

19 Source: MDPI Research, "M-PESA and Financial Inclusion in Kenya: Of Paying Comes Saving?", Hove & Dubus, 22 January 2019.

20 CNBC, "This is the real reason most Americans file for bankruptcy", Lorie Konish, 11 February 2019. See also: American Journal of Public Health (Nov 2018: https://ajph.aphapublications.org/doi/10.2105/AJPH.2018.304901).

21 Sources: National Institutes of Health and Harvard School of Medicine (2009 study): https://news.harvard.edu/gazette/story/2009/09/new-study-finds-45000-deaths-annually-linked-to-lack-of-health-coverage.

22 Source: Walmart: see https://one.walmart.com/.

23 Source: Reuters "Half of Walmart's workforce are part-time workers", 25 May 2018.

24 Source: Lancet, Volume 391, Issue 10136, 2 June 2018.

25 Source: OECD Health Statistics – https://www.oecd-ilibrary.org/social-issues-migration-health/data/oecd-health-statistics_health-data-en.

26 Source: Commonwealth Fund report (2017), E. C. Schneider, D. O. Sarnak, D. Squires, A. Shah, and M. M. Doty, Mirror 2017: International Comparison Reflects Flaws and Opportunities for Better U.S. Health Care.

27 *Augmented: Life in the Smart Lane* by Brett King, Alex Lightman, Andy Lark and JP Rangaswarmi (2015).

28 Source: AJEM, "On-demand synchronous audio video telemedicine visits are cost effective", Nord et al, August 2018/November 2020.

29 See: *The Journal of American Medical Association*, "Expenditures and Health Care Utilization Among Adults With Newly Diagnosed Low Back and Lower Extremity Pain", 10 May 2019.

30 Alliance for Regenerative Medicine Estimates.

31 Source: Optum, "How AI can help reduce $200B in annual waste", https://www.optum.com/business/resources/ai-in-healthcare/artificial-intelligence-reduces-waste-health-care-costs.html.

32 Various: PWC, McKinsey, UIPath, CiGen, KPMG research. See also HealthAffairs, "How Administrative Spending Contributes To Excess US Health Spending": https://www.healthaffairs.org/do/10.1377/hblog20200218.375060.

33 See: MDPI, "Estimating Cost Savings from Early Cancer Diagnosis" Kakushadze et al (4 September 2017) $26Bn a year in savings from early cancer detection alone.

CHAPTER 7

REVOLUTION RISK MITIGATION

"Poverty is the parent of revolution and crime."
—Aristotle

From Hong Kong to Iraq to Chile, from New York and London, Moscow to Sydney, protesters around the world have taken to the streets to rally against their governments in recent years. Is this a trend or is it anomalous? While we've had some big protests throughout history, public demonstrations are mostly a modern phenomenon.

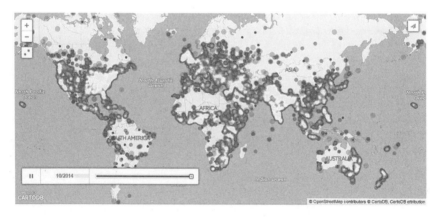

Figure 1: Intensity of protests around the world in 2014. (Source: GDELT Project)

The revolution will be tweeted

The Durants claimed that in the past, when we have seen the sort of financial inequality observed globally today, there is generally only one of two outcomes—revolution that forces political change or legislative responses that anticipate revolution and move to redistribute wealth before that happens. Today, however, we see revolution fomented in 160 characters on Twitter, with apps that use encryption to organize protests and live video streams that bring us into the moment, supercharging the forces that Durant observed.

The Arab Spring, contrary to popular belief, didn't start in Arabia nor in Egypt, but in Tunisia, bordering the Mediterranean Sea and the Sahara Desert. Tunisia experienced a series of conflicts during the period leading up to the protests that followed in Algeria, Jordan, Egypt, and Yemen, which became known as the Arab Spring. The protests all shared a sort of chaotic collectivism. There was no centralized leadership to the movement, just general broad dissatisfaction. The early protests started as rallies, sit-ins and strikes, and then descended into violence, deaths, injuries, and arrests.

The rapid escalation of early protests in Sidi Bouzid, a city in Tunisia, was centered around the death of Tarek el-Tayeb Mohamed Bouazizi, a street vendor who set himself on fire on 17 December 2010. Bouazizi, known as "Basboosa", appears to have been targeted and mistreated by local police and municipal officials over many years. The day before he died, he had been harassed by police looking for a bribe, who suggested he didn't have a permit—although it wasn't legally required. Minutes later a 45-year-old female municipal representative, Faida Hamdi, appeared and confiscated his produce, tossed aside his street cart, slapped him in the face and hurled abuse at him because he refused to pay a 'fine'. He had incurred $200 in debt the day before (equivalent to a month's wages) to buy the produce that was then seized by the police. He immediately went to the city's governor's office to complain and to ask for his goods to be returned. When he was refused an audience with senior officials, he was quoted as shouting "If you don't see me, I'll burn myself."

He left, bought a can of gasoline, returned to the governor's office, stood in the middle of traffic outside the front door and shouted: "How do you expect me to make a living?!" He doused himself in fuel, lit a match and set himself on fire. The time was 11:30am. Exactly an hour after his altercation with the police.

Within hours protests had begun. The Arab Spring spread from Tunisia to Libya, Egypt, Yemen, Syria, and Bahrain, each experiencing uprisings and mass protests against their governments. Additional street demonstrations sprung up in Morocco, Iraq, Algeria, Iran, Lebanon, Jordan, Kuwait, Oman, and Sudan, continuing well into 2012. During the peak of the Egyptian Revolution of 2011, the internet and social media played a huge role in facilitating the spread of information. Hosni Mubarak, the president and ruling authority in Egypt for 30 years, was so threatened by the power of the internet and social media in mobilizing people against his rule, that the government shut down the internet.

Within less than a year these protests had spread globally, and by September 2011 the Occupy Wall Street movement had taken up residence at Zuccotti Park in New York City. While the new residents of Zuccotti Park had differing agendas, they all agreed that social and economic inequality had reached unacceptable levels. Sixty-four percent of the Occupy protesters in the US were under the age of 35[1], with 26.7% enrolled in school or college. While conservative news outlets identified the youth as disgruntled unhinged college students, the truth is that the majority were working professionals who had been dramatically affected by the 2008 global financial crisis. But they were clearly angry, and scared.

In Hong Kong, protests had an even greater edge of anger and dismay from the youth there[2]. While a proposed extradition law was the match that lit the conflagration, the fuel was a decades-long acceleration of inequality and cost of living increases. For a decade Hong Kong's property prices have been the most expensive in the world. The average expat living in Hong Kong pays over US$10,000 per month for a moderately-sized apartment. The average local pays roughly US$2,200 for a one-bedroom apartment. For what it costs to buy a small 37.1 square metre (400 sq ft) apartment in Hong Kong[3], you could buy a chateau in France or a castle in Italy. In Sydney and Tokyo, two very expensive cities for real estate, you'd still get at least a two-bedroom apartment twice the size for the same budget. Presently, a child born in this reclaimed Chinese territory is unlikely to have the opportunity to own their own home.

Until the successful writing into law of the contentious extradition bill in Hong Kong, protests had been running uninterrupted for more than a year, commencing back in March 2019. More than 4,000 people were arrested throughout this period, and during the worst clashes of the protests, Hong Kong police reportedly fired more than 1,500 rounds of tear gas in a single day. Originally these protests were focused around the extradition bill pursued by the Hong Kong Legislative Council, under the direction of the mainland government. On 9 June 2019, more than one million protestors (about a seventh of the population) turned out against the proposed bill. The mob escalated with bricks, bottles, and umbrellas thrown at police, who retaliated with pepper spray, batons, and tear gas. The aggressive police response provoked two million protesters to return to the streets just one week later.

As protests increased, the Legislative Council headquarters were stormed, the airport was shut down, the Chief Executive of Hong Kong was forced to withdraw the bill (albeit temporarily), universities became battlegrounds between students and police, and protestors were shot. On the 24 November 2019, District Council elections were held, widely seen as a referendum on the future of Hong Kong. Ninety percent of seats went to pro-democracy anti-Beijing candidates.

Two days after the election, Beijing was unmoved. "Whatever happens, Hong Kong is always a part of China and any attempts to create chaos in Hong Kong or to jeopardize its prosperity and stability will not be successful", proclaimed the Chinese foreign minister, Wang Yi, at a press conference during a state visit to Japan. Social media has been a central part of these protests. On 21 May 2021, the Article 23 anti-sedition provision became law in Hong Kong, despite the objections of significant portions of the Hong Kong population.

On 23 September in sunny New York City, the United Nations General Assembly met and hosted the Climate Action Summit. The highlight of the day, according to the press covering the event, was a speech from a young Swedish climate activist, Greta Thunberg. After a passionate and dramatic performance, social media exploded around her speech and the so-called #howdareyou movement was born.

Prior to the Climate Action Summit, Greta had 4.5 million followers on Instagram. That has since more than doubled, and it likely won't be long before her follower count (Facebook, Twitter and Instagram) will exceed 20 million. But more importantly, the speech focused the activity of climate campaigners on social media, mostly via Instagram and Twitter.

Long-term presidents left office after successful protests in Sudan, Algeria, and Bolivia in 2019, while in Lebanon and Iraq rulers resigned. Violent unrest in Iran, India, and Hong Kong continued well into December and spilled over into early 2020.

On 6 January 2021, supporters of President Trump stormed the US Capitol Building demanding that the results of the election be overturned, that Vice-President Mike Pence be hanged for treason for not supporting Trump's call to stop the electoral college voting session, and the capture of Democratic members Nancy Pelosi and Chuck Schumer for their crimes against the nation. The result was five deaths, 138 police officers injured, and at least another 15 who were hospitalized, some suffered severe injuries. One officer lost an eye, another had broken ribs and two with smashed vertebrae. Other officers suffered brain injuries from being struck with lead pipes and other weapons. Damage to the Capitol building exceeded $30 million.

Over the last two decades we've seen a global surge in activism and protest movements after a period of relatively stable geopolitics. Based on data from the GDELT Project[4] and ForeignPolicy.com, the Arab Spring appears to have been responsible for launching a 25% increase in protest activity around the world. This elevated level of protests appeared to be stabilizing mid-decade, but spiked again around the populist and climate movements, indicating that citizen protests are playing a larger role in global politics than ever before, powered by access to social media and internet-based news outlets.

Many have even argued that the populist movement itself has largely been made up of protest votes against traditional political ideologies and constructs. However, as we're likely to see when climate change and AI make their presence increasingly felt, this mobilization will increasingly be based on anger at political systems that have failed their citizens, and leaders who have failed to act in the interests of the greater good.

This epidemic of anger that we see today is definitely global, and while corruption, racism, anti-elitism and other local issues act as triggers, the movements we see today are about calls for a more representative society and the need for social and economic freedoms across economic classes[5].

Over the last 20 years protests have more than doubled in frequency[6] and have increased roughly 1000% in terms of total crowd size and participation. Twenty-first century protests are certainly much more global in nature and have seen more event triggers than, say, during the 1960s, when the Vietnam War and the US civil rights issues resulted in widespread civil action. While not all protests over the last two decades were anti-government in nature, anti-government sentiment was responsible for an annual average increase in protests of 11.5%. Viewed in this broader context, the Arab Spring and Occupy movements were not isolated movements, they were the start of a more amplified use of protests as a tool for airing grievances and they have been enabled by social media.

The risk for society today is that these protests resemble more closely historical periods of revolutionary upheaval, not just simple political change. It can also be argued that even with the decades-long increase in mass protests around the planet, we've seen very little real effect on policy. This leads to a greater likelihood that these protests will build in frequency and support until change is forced on the political system. Trump populism and the UK Brexit are examples of where this has happened.

Revolutionary change is not new to humanity, but we've already seen how the Occupy movement, then other populist movements, were able to spread much faster and more dramatically than protests we saw in the 1960s and 1970s. Technology is increasingly a factor in mobilizing the disaffected masses, and as Millennials start to enter the political sphere, they will use technologies like social media, influencers and viral elements to gain groundswell support for policy change. But is revolutionary upheaval even remotely possible in this day and age? Isn't it more of a historical artifact?

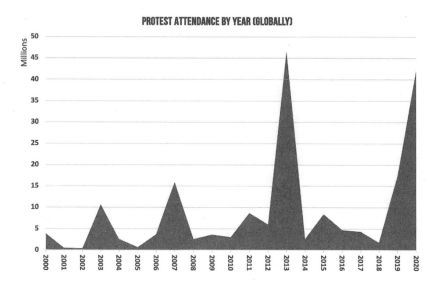

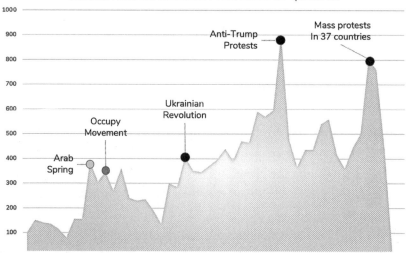

Figure 2: Protests have increased in size and frequency in recent decades.
(Source: GDELT, Wikipedia, various)

Let's look at what has led to revolutions in the past.

"Revolutions are complex processes that emerge from the
social order becoming frayed in many areas at once. There

> **are five elements that create an unstable social equilibrium:**
> **economic or fiscal strain, alienation and opposition**
> **among the elites, widespread popular anger at injustice, a**
> **persuasive shared narrative of resistance, and favorable**
> **international relations. Revolutions have both structural and**
> **transient causes; structural causes are long-term and large-**
> **scale trends that undermine existing social institutions**
> **and relationships, and transient causes are contingent**
> **events, or actions by particular individuals or groups, that**
> **reveal the impact of longer term trends and often galvanize**
> **revolutionary oppositions to take further action."**
> **—What causes Revolutions, Jack A. Goldstone (2013)**

He said it better than we could. Revolutions historically have shared the following broad criteria in the lead up to dramatic political change:

1. Extreme economic inequality or economic strain
2. Dissatisfaction aimed at the wealthy and disagreement between the elites
3. Rising populist anger, particularly at injustice
4. Growing resistance to the status quo
5. International connectedness and cooperation

If you're ticking off the boxes, yes, we are definitely in trouble.

The first two are just economic uncertainty. Addressing anger at injustice requires long-term social reform that should happen as societies mature and evolve. International connectedness will be core to dealing with climate change impact, too.

Inequality and dissatisfaction with the political elite remains a cause of division, along with increasingly polarized political ideologies, especially when it comes to potential solutions to economic problems. At some point, however, the crowd will realize that calling snap elections and attempts at creating national trade biases just hasn't worked. That economic growth, if it exists, isn't translating into real wage growth or better quality of life for most; the middle class is continuing to shrink; the number of poor or disaffected is growing; and the economic effects of the GFC and the COVID-19 pandemic continue to linger.

The risk of this turning nasty out of acute frustration is very real. We see how the echo chamber and feedback loops on social media have already created ugly, angry, divisive influences threatening to tear us apart, and how protests are increasing in volume and frequency. We've seen this frustration mobilized into mob action, with governments forced to resign in recent years. Is it really unthinkable that this could happen in more advanced democracies? The attack on the US Capitol building proves that the escalation in protests globally will continue to be a problem for even the most 'democratic' of societies.

Now let's throw in large-scale unemployment due to increasing levels of automation, the effects of climate change with ballooning numbers of climate refugees, failing crops and farmland, annual cycles of 100-year type floods in coastal cities, annual cycles of wildfires and bushfires, the loss of major ecologies (coral reefs and rainforests), increasing food scarcity, and so forth. We must recognize that this trend of global protests is going to get worse. A lot worse.

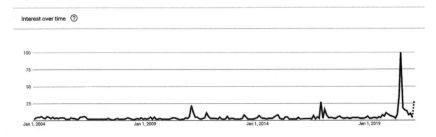

Figure 3: Google searches on "Protests" showing a significant spike more recently.
(Source: Google Trends)

The only way to ensure that revolution doesn't happen is to address some of the more fundamental problems associated with this epidemic of anger powered by social media mobilization and fake news. We need to rapidly reduce inequality, ensure social inclusiveness, and develop strategies for the increasing impact of automation on employment. We need to address the cause of broad citizen grievances, not just put a bandaid on it each election cycle. The gap between policy, demands, and the ability to govern appear to be growing, not shrinking.

We either mitigate these risks, or some form of revolution may just be the next chapter we write for the 21st century history books.

The case for UBI: Mitigating techno-unemployment and inequality

Where we come from in Australia, someone who doesn't work and survives on government support is labeled a "Dole[7] Bludger". In the US you might hear "Welfare Queen", welfare parasite, mooch, leech, freeloaders and welfarians to describe someone who is "living off the government tit". In the UK, you get "lazy scroungers" who are "on the Dole", but in Scotland someone on the dole is referred to as "going on Pogey". Regardless, there's generally a strong negative connotation to receiving benefits, along with a social stigma attached to being unemployed.

This makes the concept of a universal benefit like UBI (Universal Basic Income), or some form of basic asset support, seemingly a tough concept to sell. That is, until you're unemployed, an AI has taken your job and you don't have enough cash to buy the next meal or pay the rent. Or until a pandemic comes along.

In a McKinsey Global Institute report of 2017, it was estimated that as many as 73 million jobs would be lost by 2030; the report also posited that about 20 million of those lost jobs could be mitigated with retraining and reskilling. A report by the Brookings Institute in 2019[8] showed that at least 25% of jobs in the US are likely to be hit by automation, especially "boring and repetitive" ones. PricewaterhouseCoopers estimated 38% of all jobs over the next 15 years would be disrupted by automation.

The University of Oxford conducted even more extensive research and came to the conclusion that markets like the US stood to lose 47% of their jobs to automation. The Oxford study completed in 2013 and revised in 2018, examined over 700 occupations that make up about 97% of the US workforce[9]. Included in the panel of experts were specialists in Machine

Learning (ML) that assessed the ability of algorithms to impact those 700 occupations. Five years later, those same researchers assessed the likelihood of disruption by AI as even more acute as improvements in Machine and Deep Learning are realized.

With such potential dislocation of traditional occupations over the next 20 years, along with stagnant wages and a steadily rising cost of living, the need for a cushion or social safety net is imperative. We can either wait until large-scale unemployment threatens revolution, or we can model potential systems to mitigate worst-case impact.

Will capitalism fix labour displacement on its own?

Diehard capitalists would argue that capitalism can fix most of the problems we face. Not only hasn't capitalism addressed inequality, but capital markets driven by shareholder return are simply not incentivized to solve social problems such as those that have led to these protests we see around the world, and the free market has had no problem with policies that have accelerated inequality globally.

Greta Thunberg ✔ ...
@GretaThunberg

Isn't it funny that the same people who are calling us naive for believing change is possible are the ones who believe fossil fuel companies, big polluters and oil producing nations will "fix" the climate- and ecological crisis with their vague, distant "net zero" commitments...?

6:21 PM · Mar 24, 2021 · Twitter Web App

2,065 Retweets   **62** Quote Tweets   **12.2K** Likes

Figure 4: It seems terribly naive to think capitalism will address climate change. (Source: Twitter)

The biggest problem for capitalists arguing that inequality and climate change will be addressed by the free market is that capitalism has done a poor job of managing those issues thus far. Sure, if renewable energy were cheaper than fossil fuels, then investors might invest in retooling energy infrastructure, but we still see market resistance to this in the US, India, Russia, and Australia, despite solar and wind being "the cheapest form of energy in history"[10]. Those same industry players did very little over the last century when seemingly aware that pollution was already killing millions annually. Ambient air pollution today accounts for between seven and nine million deaths annually due to stroke, heart disease, lung cancer, and acute and chronic respiratory diseases[11]. That's more than those killed annually by smoking and COVID-19 combined.

It could be reasonably argued that fossil fuel corporations (Big Oil, Big Coal, etc.) and the markets have fought every step of the way against broader social needs in favour of continued fossil fuel returns and profits. From funding medical professionals who dismissed black lung[12], to funding climate denial think tanks and PR efforts to undermine legitimate climate science[13], the free market has rejected interventions that would dramatically restructure energy markets (unless permitted by those same markets). In respect to pollution and climate change specifically, this makes very little sense when you step back from the market and look at sensible tradeoffs between energy and the health needs of the planet. If your only frame of reference is GDP growth or shareholder returns, then can you really argue that nine million deaths annually from pollution-related conditions are acceptable costs of increasing shareholder returns? Let's do the math.

The global cost of air pollution caused by fossil fuels is estimated to be $8 billion a day, or roughly $2.9 trillion per year or equivalent to 3.3% of the entire world's economic output[14]. China, the United States, and India are responsible for $900 billion, $600 billion and $150 billion per year respectively. The Centre for Research on Energy and Clean Air (CREA) report showed particulate air pollutants account for around 4.5 million premature deaths each year, including 1.8 million in China and a million in India. This aligns with the WHO's estimates of previously around 4.2 million deaths.

For $2.9 trillion per year in negative impact, the industry returns $200–260 billion in profits on $1.2 trillion in annual revenues—that's excluding global subsidies, which totalled $5.2 trillion in 2017. This doesn't include the trillions of dollars of negative climate impact that these companies are responsible for. So $2.9 trillion in health impact, plus $5.2 trillion in taxpayer subsidies, against, say, $250 billion in annualized profits. This leaves us with a global opportunity cost of $8 trillion in annual write-offs that society carries as a net loss from the fossil fuel industry.

The math just doesn't add up. Based on the cost to society, and the long-term impact on the market, we should be mandating a shift away from fossil fuel usage as rapidly as we can, not allowing the free market to determine when this is appropriate as measured by profitability.

Likewise, as robots and automation make companies more productive, markets won't complain that dividends and earnings are increasing exponentially while human workers are being laid off by the millions. Can you imagine the stock market discounting Uber's stock price, for example, when they announce the deployment of their long-awaited self-driving fleet and the fact that they're phasing out human drivers? Would you expect Amazon's share price to dip if Bezos announced their warehouses were going fully autonomous? If Apple decides to bring their factories back to the US, but those factories are essentially robotic factories that don't dramatically increase employment, would the market or the government push-back, demanding more human jobs?

Based on precedents, there is absolutely no evidence that the free market will correct itself to adjust for changing employment conditions based on high levels of automation that displaces human labour, nor for reducing climate impact today or improving the health of individual citizens.

If large-scale unemployment happens, how do we pay for UBI?

Bill Gates says we're going to have to tax robots to fund Universal Basic Income (UBI), but he's not the only one. In fact, it's becoming increasingly normal for entrepreneurs to argue in favour of UBI. Andrew Yang, Democratic presidential candidate for the 2020 Election, ran largely on the platform of UBI, or as he called it "Freedom Dividends".

The concept of universal income didn't originate with 2020 US Democratic candidates or modern-day entrepreneurs. Martin Luther King, Jr., former President Richard Nixon and economist Milton Friedman are all examples of those who championed versions of UBI in the past.

Some of the entrepreneurs who have endorsed some form of basic income to solve inequality and employment disruption are:

- Elon Musk (Telsa/SpaceX)—"It's going to be necessary."[15]
- Jeff Bezos (Amazon)—"[Bezos] has given up on the notion that our economy, or [Amazon], can support that pillar of American identity: a well-paying job."[16]
- Jack Ma (Alibaba/Ant Financial)—"Universal basic income should also play an important role…it will dramatically reduce poverty and unemployment."[17]
- Mark Zuckerberg (Facebook)—"We should explore ideas like universal basic income to give everyone a cushion to try new things."[18]
- Jack Dorsey (Twitter)—"…fascinated by the idea of UBI as perhaps the most effective way of redistributing his wealth."[19]
- Pierre Omidyar (eBay)—Donated $493k to a UBI experiment in Africa called GiveDirectly.[20]
- Ray Kurzweil (Google/Singularity)—"In the future you'll do something you enjoy."[21]
- Richard Branson (Virgin Group)—"A basic income should be introduced in Europe and in America."[22]
- Tim O'Reilly (O'Reilly Media)—UBI will "reshape the definition of work itself, and give people more flexibility to do the things that feel most personally fulfilling."[23]
- Chris Hughes (Facebook)—"Rather than try to restructure our economy so it looks like the 1950s… we need to consider what systems we'll need to create if millions of jobs disappear."[24]
- Sam Altman (Y Combinator)—Pioneered trial giving $2,000 per month to LMI segment.[25]
- Stewart Butterfield (Slack)—"…doesn't have to be much, but giving people even a very small safety net would unlock a huge amount of entrepreneurialism."
- Andrew Ng (Coursera/Baidu)—"More than ever, we need basic income to limit everyone's downside."[26]

These are all extremely intelligent, highly competitive entrepreneurs and business operators. Smart enough to own companies collectively worth many trillions of dollars, and they aren't dismissing UBI as a crazy idea put forward by Marxist economists. They are seriously considering the possibility of UBI, even endorsing the need for it. Why?

These entrepreneurs not only see the potential for massive disruption of employment because of automation, but they are also *actively building businesses* that will be less and less dependent on human workers over the coming decades. They are smart enough to realize as job losses from automation stack up, we will need a plan to placate those most affected by these changes. If not UBI, what else?

Universal Basic Income also has broad public support. A recent study by Oxford University showed that 71% of Europeans favour some sort of UBI[27], while 84% of Europeans support a mandatory minimum wage. In his 2020 Easter letter to Catholics around the world, Pope Francis even agreed it might be "time to consider a universal basic wage".

As discussed earlier, human labour has become less attractive to employers in the last decade or so because worker productivity has stagnated. Labour productivity rose at an average annual rate of only 1.1% between the fourth quarter of 2007 (when the Global Financial Crisis began) and the third quarter of 2016. In fact, most of the productivity gains of the period 1990–2007 were attributable to IT spend and technology improvements, and not human labour productivity. Thus, it is reasonable to assume companies will continue to seek out efficiency gains using technology and not human labour process improvements.

In the United States today, payroll taxes account for more than one third of the annual federal budget, or approximately $1.3 trillion. In 2019–20, the value of HMRC tax receipts for the United Kingdom amounted to approximately £634.64 billion ($890 billion). If half of the workforce is displaced by robots, tax revenues will take a major hit in the world's leading economies. To be able to fund a form of Universal Basic Income without new sources of revenue, governments are going to run into serious budgetary constraints. There's no way that we avoid changing the paradigm of labour correlation with income taxation in the 21st century—it is simply not sustainable.

What UBI trials tell us

In Spain, the Minimum Living Income (*Ingreso Mínimo Vital* or IMV) is the first national minimum income scheme to be implemented in Europe. Although its approval was accelerated to tackle the social and economic consequences of the coronavirus pandemic, the IMV had already been approved in the coalition government's agreement between the Spanish socialist party, PSOE, and Podemos. In May 2020, a survey to assess the introduction of a UBI in Spain was conducted and revealed that 56% of Spaniards were in favour of it. For those earning less than €1000 per month, 67% were pro-UBI.

The national IMV program is a non-contributory cash benefit. The benefit is not attached to employment history and is primarily addressed to households who are below a recognized income threshold. This minimum income is still below the poverty line, but is designed to support nearly one million households and so far, is aimed at the poorest 2.3 million people (out of a population of 47 million). The Spanish social security institution has calculated that around 550,000 households in Spain remain in extreme poverty, that is, with income below €230 per month ($275 per month). While the IMV may very well eventually create a minimum income scheme for the entire Spanish state, it is currently a basic income for the poorest people in the country. Will it lead to a UBI for all Spaniards? Only time will tell. For those that have access to the program today, it has been life changing.

In California back in 2019, the City of Stockton launched a small-scale UBI trial. Facebook co-founder Chris Hughes was the tech entrepreneur responsible for the idea, and he and his Silicon Valley pals funded the trial— no tax dollars were used. The program sent monthly payments of $500 to 125 randomly selected individuals with incomes lower than $46,033 per year (the city's median income level). There were no strings attached, no qualifying requirements except for the lower-than-average income.

When the trial started, only 28% of those selected had full-time employment. By the end of the first year of the program, more than 40% now had a job. Had the program mirrored Stockton's general employment increase during the same period it would have only reached 24%. At the conclusion of the trial, 62% of the group said they were now paying down debt regularly, whereas this was just 50% the year before.

Participants received their basic income via a debit card. Transaction analysis showed that the number one spending category each month was on food, second highest was general shopping at merchants like WalMart and Target, which also sell groceries of course. Then it was utilities, auto payments and travel. Last was tobacco and alcohol, which accounted for less than one percent of spend.

Finland is another European country that is home to one of the world's largest UBI trials. Finland's two-year UBI pilot, which ran in 2017 and 2018, paid 2,000 randomly selected unemployed people across the country a regular monthly income of €560. This was then compared with the 173,000 people on Finland's standard unemployment benefits program. The study carried out by the Finnish government showed mixed outcomes for both supporters and critics of UBI. But it did result in significant improvement in wellbeing over standard unemployment programs.

> *"The basic income recipients were more satisfied with their lives and experienced less mental strain than the control group... They also had a more positive perception of their economic welfare."*
> **—Helsinki University Report on Finnish UBI program**[28]

Helsinki University researchers, who conducted 81 in-depth interviews with participants in the scheme, concluded that while there was significant diversity in their experiences, recipients were generally more satisfied with their lives and experienced less mental strain, depression, sadness and loneliness than the control group. Ironically, access to UBI had a net stimulus effect on employment, particularly in certain categories, such as families with children, going against the assertion that UBI would be a disincentive to work. Participants also tended to score better on other measures of wellbeing, including greater feelings of autonomy, financial security and confidence in the future.

The scheme also gave some participants "the possibility to try and live their dreams", said Professor Helena Blomberg-Kroll, who led the study. "Freelancers and artists and entrepreneurs had more positive views on the effects of the basic income, which some felt had created opportunities for them to start businesses." Others used UBI to get more involved in society by taking on voluntary community-based work.

In short, UBI eliminated economic insecurity for those who participated, giving them choices beyond simple subsistence based on minimum living wages. Given the economic uncertainty at the heart of climate change, automation, and rolling pandemics, UBI seems like the most humane solution available on a medium-term basis—but only if you want happy and fulfilled citizens. The only question remains is how can we pay for it?

Figure 5: Locations where UBI trials have been successfully executed. (Source: various)

Paying for UBI

There are four potential ways we think we might be able to fund Universal Basic Income beyond just provision of UBI from the usual government budgetary approaches.

Figure 6: Paying for Universal Basic Income.

Funded by Big Tech?

In Chapter 3, where we examined the increase in the value of top technology companies to potentially tens of trillions of dollars and their founders perhaps approaching trillionaire status, you can imagine these companies being targets of bad PR around their negative impact on employment through technology automation. Even now we see the likes of Google and Apple starting to invest in public-focused training, supporting UBI trials and lobbying for UBI, improving access to early STEM development and so forth. This is clear evidence that they realize they need to stay on the good side of potential end-users of their tech, some of whom might be adversely affected by the implementation of AI and technology disruption.

When climate impact starts to really hit, we can also expect major players like Bill and Melinda Gates, Warren Buffett, Elon Musk, Jack Ma,

MacKenzie Scott, and Jeff Bezos to put hundreds of billions of dollars into climate mitigation programs. By 2050, climate mitigation programs will be one of the biggest net employment growth areas.

Elon Musk announced $100 million sponsorship of a Gigaton Carbon Removal program commencing on 22 April 2021 (Earth Day) and running for four years. The prize includes 25 $200,000 student scholarships to participating student teams, along with $50-, $20- and $10-million purses provided to the top three technology demonstrators at scale.

Figure 7: Musk has already committed $100 million to carbon sequestration development. (Image: XPrize website)

Bill and Melinda Gates have previously asserted that one third of Americans will need to switch careers by 2030 alone, which is backed up by McKinsey research. Speaking at an event in 2018, Melinda Gates stated that emerging sectors such as robotics, machine learning, and AI are already shaping the types of jobs that will be available through this decade. She maintains that gaining expertise in new industries will be a key deciding factor in our ability to stay employed in the future. Their foundation has already dispensed $54.8 billion of grants that have created jobs in 135 countries.

Wealth generated by AI

Here's a novel concept. What if anyone that loses a job to AI just continues to get a paycheck for the job they lost, but the AI actually does the work? Would anyone object to that?

OpenAI's Sam Altman claims that we're not going to have to worry about how to pay for Universal Basic Income. He argues that AI is going to make so much money for the global economy, that paying people a basic income will not only become an easy problem to solve, but that it will also allow us to "align incentives" around the most critical problems humanity will face, like climate. But losing jobs to AI should not be something we need to worry about.

Altman says that by 2030, AI will be able to fund $13,500 per year for every adult living in the United States. But he also says that this requires a fundamental shift in government policy for this to work. To get this number, Altman estimated that by 2030 there would be $50 trillion worth of value in leading US companies (calculated by market capitalization) and $30 trillion worth of privately-held land in the US. Altman argues for the establishment of an American Equity Fund that taxes sufficiently large companies 2.5% of their market value (in the form of equity), and 2.5% of the value of all land (in the form of dollars). Companies worth $1 billion or more utilizing AI would also contribute into the fund.

Figure 8: OpenAI co-founder Sam Altman argues AI will pay for UBI.

As the pace of development accelerates, Altman argues AI "will create phenomenal wealth" but at the same time the price of labour "will fall towards zero... It sounds utopian, but it's something technology can deliver (and in some cases already has). Imagine a world where, for decades, everything—housing, education, food, clothing, etc—became half as expensive every two years."

This is technosocialism, clearly.

Central bank digital currencies

Central bank digital currencies (CBDCs) are gaining a great deal of traction and interest globally today. Many are debating whether China's digital yuan might disrupt the dominance of the almighty US petrodollar. Although China's CBDC is designed for an entirely different purpose, it has demonstrated the success of linking consumption directly to digital currencies.

China's CBDC has already created four government-run mobile wallet types, aided in implementation by Alipay and WeChat Pay, and with increasing transaction limits and numerous phases of deployment. Four different large trials have taken place across two phases thus far. CBDC trials have taken place in four Chinese cities: Shenzhen, Suzhou, Chengdu and Xiong'an. The wallets are largely anonymous and allow a maximum of 500 yuan ($77) per payment, a daily limit of 1,000 yuan ($154) and 10,000 yuan ($1536) per month. Shenzhen, the southernmost city bordering Hong Kong, has already entered its third eCNY CBDC trial. A hundred thousand people were initially given 200 yuan each in the latest trial.

Digital yuan wallets have been tested for hiring bicycles by Meituan and Qingju Bicycle share services, and more recently, the super app and ride-sharing service Didi Chuxing now supports eYuan payments for ordering a taxi, food delivery, sending parcels, booking theatre tickets, travel, etc. E-commerce player JD.com, along with its group-buying app Jingxi, and video sharing service Bilibili, are also actively participating.

The government of Macau injected stimulus payments into their local economy by distributing contactless Macau Pass smartcards to residents that came pre-loaded with a credit of 3,000 patacas ($377) that had to be spent between May and July 2020. An additional 5,000 patacas payments ($629) were issued for the period of August through December. The cards could not be used for spending in casinos, pawnshops or financial institutions. They could not be used to buy airline or ferry tickets, tourism outside of Macau, or to pay utility bills. UBI-focused CBDCs could easily be encoded as smart money that could only be used for the purchase of staples like rent, food, medical expenses, clothing and education. However, over time UBI is likely to become a base income for lower-middle income households impacted by unemployment, and not be restricted in this way.

CBDCs could in effect be a secondary money supply that is purpose-built for UBI. It will stimulate consumption, creating jobs and economic activity, but won't necessarily affect broader trade and commerce. CBDCs could effectively create closed-loop ecosystems where UBI-based currencies could be used for your daily needs, but not for general consumption outside of that.

Climate mitigation and techno-reform

As mentioned earlier, we propose a global forgiveness of national debt, where that national debt is committed to climate mitigation efforts over the next 30 years (also see Chapter 9). This serves two broad purposes: one, committing the world to serious climate action; and two, providing national infrastructure projects that will generate jobs for those displaced by automation and changing skill sets. However, it is not inconceivable that we could choose to mobilize massive national workforce programs around climate change mitigation instead of just a straight UBI stimulus, or provide greater incentives for workers who choose these new climate-mitigating sectors.

The International Renewable Energy Agency predicts that by 2050 more than 42 million new energy infrastructure jobs will be created in response to massive cost benefits and energy infrastructure reform and modernization programs, which is broadly classified as energy transition.

By the end of the 21st century, we predict there will not be a single operating coal, gas, or nuclear energy plant on the planet (there might be some nuclear energy plants on Mars and the Moon, but even there *in situ* solar will be a more likely longer-term solution). In fact, it's fairly probable that we will be 100% renewable for electricity generation sometime in the 2050s. This will require the entire energy grid to be refitted for distributed solar, wind, geothermal, and hydro generation, with energy storage (battery) farms around the world to store energy for later use. Innovative technologies like molten salt batteries could provide grid-scale energy storage capability that is 5–20 times more effective than lithium-ion (Li-ion) batteries[29].

With Li-ion, Tesla was successfully able to power a 100 MW battery farm in Adelaide, South Australia. The battery farm was a result of a bet that Elon Musk made publicly on Twitter with then Prime Minister Malcom Turnbull and the now current Prime Minister Scott Morrison, who had both blamed[30] rolling blackouts in the southern state on renewable energy generation. One central argument was that battery storage could not cater for peak load demands on the network, and that only fossil fuel plants could spin up quickly enough to satisfy the sort of demand South Australia faced when storms took a heavy toll on the energy network.

Musk not only installed the 100 MW (129 MWh) Tesla giga-battery storage farm in the required 100 days, but demonstrated for the first time that a battery storage facility could replace natural gas-fired electricity plants when it comes to peak load profiles and shedding. The dedicated battery farm can power 30,000 homes for up to an hour, which relieves the burden on the grid during hot summer days when failure is most likely. This load-smoothing capability saved South Australians around A\$116 million (\$76 million) in reduced network costs in 2019 alone. The battery's introduction also slashed the cost to regulate South Australia's grid by 91%, according to Gareth Heron, the head of development for South Australia operator Neoen. This is probably why the government opted to increase the battery farm by another 50% in late 2019.

Here are a few other areas where massive investment will be required globally because of climate shift, and will in turn create entirely new growth sectors:

1. **Carbon Sequestration**—IPCC estimates are that we will need to pull somewhere between 100–1,000 gigatons of carbon dioxide from the air by 2100, and that is just at the level of reversing the last 20 years of carbon output. The Trillion Tree campaign, the Musk XPrize for Carbon sequestration, etc, are all examples of this global push.

2. **Seawall Defences**—in 2014 New York launched a $335 million seawall defence initiative[31] to protect New York City from events like Superstorm Sandy, which created $19 billion of damage to the city.

3. **Climate Resistant Infrastructure**—reinforcing existing critical infrastructure against frequent flooding will be vital in coastal cities, including power, sanitation and water infrastructure, road, bridges and public transportation systems, and hospitals and emergency services.

4. **Green Industrial Reform**—in November 2019, a Bill & Melinda Gates Foundation sponsored solar stealth startup called Heliogen was able to use directed solar to heat a solar oven to in excess of 1,000 degrees Celsius, providing the breakthrough we need to manufacture glass, cement, steel and other industrial processes that rely on fossil fuels.

5. **Relocation of Settlements and Industry**—by 2050, 95% of northern Jakarta will be underwater, so in August 2019 Indonesian President Joko Widodo announced that a $33 billion project would commence to build a new capital city in East Kalimantan province, citing flooding concerns.

6. **Sustainable Consumption**—from lab-grown foods and high-order recycling through to vegan leather based on mycelium-based leather substitutes[32], we can expect the world to see products based on non-sustainable components through a lens similar to the one we use to view blood diamonds or nuclear waste today. Companies will be rated on their ability to recycle and use sustainable materials transparently, and stock markets will include more climate-friendly ratings than just share price and net earnings.

But the bigger impact on employment, as previously discussed, will be the high-level of techno-based unemployment. While shortages will remain for technical roles in the economy, many existing jobs will be changed beyond recognition by technology. How will automation more broadly affect the role of work in mature economies? To answer that, we should really look at history.

How work is going to change for everybody

The full-time job[33] is historically an anomaly. Prior to the Industrial Age, it didn't really exist. Early industrialists, who needed to have workers on a production line simultaneously for efficiency, are most likely responsible for creating the concept of a structured working week. For the last 100 years the 40-hour-a-week job has been the centerpiece of work life simply because there was no better way for people to gather in one place at the same time to connect, collaborate and create output.

Most people will spend 90–120,000 hours working during their lifetime, the equivalent of 13–14 years of non-stop, 24/7 effort. It's reasonable that work defines the answer to "what we do" in our waking hours, and indeed, how we most commonly identify ourselves to others. Our work and identity are intertwined. It's why employment and work remain critical measures in an economy, and where people find self-worth. As automation impacts society more and more, the answer to this question must go through a fundamental change.

If we're right about automation and its impact on society, by 2050 large portions of society won't work the 9–5, 40-hour week that is common today. With basic sustenance, energy, food, healthcare and education provided as part of a competitive smart city infrastructure, you will likely be able to eke out a basic existence without working full-time. Of course, if you want to travel, buy the latest gadgets, eat at upmarket restaurants and so forth, you will probably need to contribute more than if you are happy to do without those things. The core problem is that we'll all be working less. But how much less?

If long-term trends continue at their historic rates, we can expect that most people will be working around 25 hours or less by 2050[34]. If you add this to the fact that life expectancy is estimated to rise to an average of 90–95 years in developed nations, with productive working years remaining in the 20–70 age range, this means that retirement propositions are very different in 30 years' time. For one, your savings will have to last effectively twice as long as is expected in today's retirement scenarios. Secondly, reduced working hours over your working lifetime could lead to lower take-home pay and reduced savings over time. Regardless of the effects of Artificial Intelligence and technology-based unemployment, we're still going to need increased social safety net infrastructure, particularly for older populations.

When we examine historical employment numbers, while we've generally maintained a consistent ratio of employment to population, working hours have continued to decline since the 1950s, reducing by about 25–30% in the last 50 years alone.

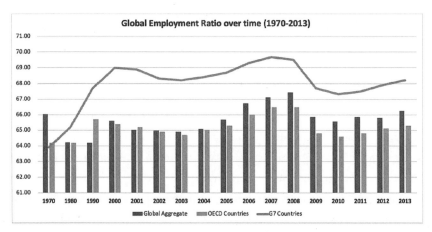

Figure 9: Ratio of global employment to population vs average working hours. (Sources: OECD, University of Groningen)

This is just based on current trends and doesn't factor in technology-based unemployment. US Bureau of Labor Statistics research shows labour force participation declining from its peak in the 1970s at 2.6% annual growth, to as low as 0.5% in the coming decade. In addition, slowing population growth will produce an ageing workforce with a doubling of the percentage of workers aged 55 and above by 2050 (from 13% to 24%[35]).

Japan raised their national retirement age to 65 in 2019. Australia and other countries are pushing out access to national pension schemes to 67 by 2023, based on an ageing population and declining pension pools. In the United States alone, the American Legislative Exchange Council estimates that unfunded liabilities based on trending pension fund returns is set to exceed $6 trillion in the near term.

At some point in time the conflicting trends of an ageing workforce dependent on continued employment, continued reduction in working hours, increased technology-based unemployment and greater dependency on savings for living well into our 90s, all come crashing together into some sort of chaotic policy nightmare. Absent some very complex planning and policy forethought, the largest and oldest demographic segments of society could be abandoned in droves by a simple failure of capitalism to anticipate this level of systemic change.

You can't look at these trends and conclude that there will be enough work for everyone. Employment, and work in general, will fail to deliver the outcomes that it must for society to function in an orderly and equitable manner unless we rethink how this should work. It doesn't even matter how prepared you are to work hard or how prepared you are simply to accept what is available. Even if you can find a job, it may not deliver enough working hours at a reasonable enough rate to give you a decent living wage, let alone when you reach retirement age. Don't forget the figures from Chapters 1 and 2, which show that in the United States a person on minimum wage working 40 hours a week today can't even afford a basic two-bedroom apartment[36].

"What do you do?"

How will we answer that question in the future?

It is likely that by 2050 the answer will be much more complex than today. Like the out-of-work Los Angeles-based professional that introduces himself as a producer, writer, director, and actor, in the future we may have a collection of pursuits that consume our time. We might have some work that augments our Universal Basic Income, we might be a lifelong student, we might be passionately engaged in some broader social activity or cause, or we might commit our life to reversing the damage done to our planet.

The role of work is set to become secondary to most of us, largely due to the abundance that AI will create, and the destruction of traditional process-based work that has dominated since the Industrial Revolution. But beyond work, what will shape the value systems of the 21st century?

Emerging value systems of the 21st century

The Great Depression of 1929 devastated the global economy. A third of all banks in the US failed. By 1933, gross domestic product (GDP) per capita in the US had fallen by 47%. Most of the world's largest economies experienced unemployment of 20–25%. In industrial and mining areas of the UK, unemployment reached almost 70%, with more than 3.5 million Brits out of work. Homelessness blew out. Housing prices plummeted 67% and international trade collapsed by 65% in the United States, while trade dropped by half in the UK. All told, it took almost 25 years for the stock market to recover.

Following the 1930s financial collapse, economic policies became much more collectively focused. Like today, economists argued passionately for various policy approaches. Most governments tended away from the conservative policies of the early 20th century, as they were seen as strongly linked to the market failures of the Great Depression. Typical strategies included efforts to keep labour costs low to encourage employment and stabilize currency value. In the US, the New Deal was established to put Americans back to work through vast public works programs for the unemployed.

The generation that went through the Great Depression was forever changed by its impact on society. Combined with the Depression of the 1930s and World War II, people became very savings-focused. It wasn't until the mid-1960s that consumerism came back into fashion. Those that went through the depression had memories of people losing their fortunes through the collapse of the banking system, and this would remain an embedded memory well into the late 20th century. Like the Great Depression, the 2008–9 Great Recession, combined with the recent pandemic, will change the behaviour of 21st century citizens, along with the impact of the internet, climate change, and things like "fake news".

By 2030, Gen Ys will constitute the largest demographic, and they'll also be setting macroeconomic policy—or at least demanding a seat at the table. The great dream of owning your own home (with or without optional white picket fence) is clearly unattainable for the majority of Gen Ys, Gen Zs, and Gen Alphas (those born since the early 2010s). If working till you retire just so you can pay off your mortgage is not going to be the lifelong mission of our children, what will it be?

Our children will be far less asset-focused during their lifetime, instead aiming to maximize their experiences, and will be much more focused on building a better future as a globally-connected species. This is a generation that has grown up seeing plenty of division around economic status, race, gender, and sexual orientation, but has played a large part in battling such discrimination. This is a generation that has grown up seeing fierce debates around immigration, privacy and welfare for older generations, but who instead is connected with friends around the world as a result of the internet, social media and gaming. They are much more collectively-oriented, concerned about the world they are being handed by their parents after centuries of environmental neglect and unbridled exploitation of resources. Their parents argue about capitalism versus socialism, and view the purpose of the economy as being to fulfill their basic needs first, before billionaires add more zeros to their bank accounts.

This generation is not excited about being the most powerful economy in the world, having the largest army or winning trade wars. They have national pride, but they see that they are part of a global, interconnected family that is dependent on each other for more than a fragile co-existence. They have been through two financial crises in the first two decades of the century, and they see the futility of economic policy when they can't afford to move out of their parent's basement or they graduate with a bachelor's degree, $100,000 in student debt, and then can't get a job. They hear economists talk about GDP growth and companies worth trillions of dollars, while they also see education, healthcare and housing markets that are collapsing under growing mismanagement, systemic bias, and a lack of development and funding.

What is different about the digitally-native generation born into the 21st century is that they largely assess the 20th century as a failure for the human species. They saw uneven wealth accumulation that resulted in the worst inequality in modern history. They saw housing values plummet during the GFC, and then they saw millions of families who couldn't make rent face eviction during the pandemic while wealthy "Boomers" made bank. But most of all, they saw their parents' generation arguing about whether climate change was even real, while the sixth great mass extinction event was accelerating[37] all around them, screaming at them for collective action.

Abraham Maslow's "Hierarchy of Needs" (1943) is a motivational theory in psychology. The model, in classic terms, is broken up into five tiers of human needs that are generally represented in a hierarchical pyramid structure. Although Maslow did provisionally add another three levels at later stages, the classic model is five-tiers, with the bottom four constituting deficiency needs and the fifth (self-actualization) being a growth need. The four deficiency needs are esteem; love and belonging; safety and security; and physiological well-being.

The deficiency needs application falls into what Maslow categorized as areas where humans classically feel deficiencies (i.e., I'm having a baby and I need a bigger house; I am hungry; I am lonely; I feel unsafe, I feel overweight, etc). Maslow argued that, before people could move up the pyramid tiers to reach their full potential, they needed to fill the gaps on deficiencies that preoccupied them. The longer you deny basic needs, he suggested, the higher the motivation to fill those needs becomes.

For the post-Gen X generations, their needs have been framed by the world they are born into. The basic physiological needs can all be met quite easily today (as long as you have access to money), so the pursuit of these basic needs as a driver have morphed due to technology adoption, internet access and greater transparency and communication, as well as greater social pressures and economic uncertainty.

Maslow's hierarchy provides a framework for personal philosophical beliefs. Maslow's framework was designed with transcendence in mind, the ability to continually better ourselves. But Maslow argued that we are constrained in terms of personal growth by having to address those deficiencies first.

The post-coronavirus generations, as they enter the era of broad climate disruption and AI-based changes to humanity, will seek to slow down asset and wealth accumulation, while searching for some stability over their future. Benjamin Mann at SnapMunk revisited Maslow's hierarchy of needs based on emerging behaviours in the Millennial segment.

Figure 10: Maslow's hierarchy rebuilt for the 21st century.
(Revisionary Maslow's Model Credit: Benjamin Mann, SnapMunk)

In this 21st century reinterpretation of Maslow's framework, the priorities of Millennials are most definitely changing. Maslow talked about "Peak Experiences" in his later work on the hierarchical model, meaning the higher up the pyramid you go, the more deeply satisfying experiences you will find. The original model separated psychological needs and safety needs, but for the Millennial, that distinction really doesn't exist anymore. Safety is a precursor to being happy, thus it fits with the fundamental needs of food, warmth and shelter—this instantiates itself as stability. In today's environment with peak "economic uncertainty", Millennials are looking for a way to eliminate that anxiety around money and wealth. The next tier refers to both digital and social accessibility. Millennials are uniquely connected to the world through communication tools that didn't exist in Maslow's era. But very visible to them is inequality, bias and lack of access.

Millennials' view of **health** is much more than simply whether you are sick or not, it is more concerned with mental health and the ability to leverage resources for personal progress and growth. Realistically, limits on access to basic services often restrict growth opportunities.

The social world emerging on top of the internet has created a feedback loop for younger Millennials, where their own self-esteem is somewhat linked to numbers of followers on TikTok or Instagram success. This is where **love** and broader social recognition is important for digital natives; the other side of the coin is that this can become a powerful motivator for growth.

That leads to the final tier, which is **legacy**. However, we're not talking about the sort of legacy when a president is coming up to the end of their term; it's more a collective legacy of our species—the ultimate in transcendence beyond the self. It's quite the aspiration this generation has, but their personal experiences have shown them humanity needs to work more effectively together to leave the planet a better place, and humanity better off.

This brings us to the heart of the major economic behavioural shifts emerging for the 21st century.

Experiences over "things"

For teenagers who obsess over having the latest iPhone or game console, you'd think the need to acquire the latest and greatest thing would be a core driver—but no, that's not the way it works. A competent smartphone is like having enough food to eat, or a warm bed. It's not optional, it is how a Millennial connects and thrives. But for big-ticket assets like cars or houses, they're not as excited or motivated.

A study called the Harris Report[38] showed that 72% of Millennials prioritize having memorable experiences over buying assets or something desirable. When Millennials talk about FOMO (Fear of Missing Out), they're not talking about missing out on a new PlayStation PS5—they're talking about missing out on an experience!

This has in turn fuelled growth of services around the "Experience Economy". Uber, AirBnB, WeWork winning over long-time retailers who are today declaring Chapter 11 bankruptcy. It is clear this is a macroeconomic shift. If you look back at high-growth periods in China over the last two decades, and the US in the 1950s and 1960s, the housing and real estate boom was a huge driver of economic growth and consumption. But technology-led economies with high levels of automation definitely lend itself to the experience economy.

We can see that driving humanity's future over the next century will be stabilizing life, reversing climate change and undoing inequality, but staying centered and motivated, Millennials will then pursue life experiences instead of putting down roots. Housing will be a shared societal solution at some point for anyone who chooses; we will focus on long-term utilization of consumed goods, reusability and recyclability, rather than simply wanting the annual iPhone update. All of this leads to less consumption of goods, and more consumption of services. This is a core shift in economic behaviour, one that this generation will lead. It also leads, naturally, to a much greater global coherence, as many more people embrace multiculturalism, the rich culture of differences and the truly connected existence we share. Together we will fix the world, and together we will experience the world. That is our mission.

But if that is the case, how should we measure economic growth? With a focus on a reusable, sustainable economy, consumption may very well fall out of favour. Investment in infrastructure and basic needs will grow, but that isn't as scalable in terms of profitability as global smartphone sales or monthly active users. Surpluses and deficits are replaced by carbon neutrality, helping reverse inequality and climate damage. The economy will be at the same time very introspective and very globally connected. It's sort of beautiful, when you think about it. Everything that has allowed us to create commerce and markets thus far can now be utilized for the future betterment of humankind.

The 21st century economy will be one that attempts to prioritize the basic needs of citizens over the simple creation of wealth.

Endnotes

1   Demographics of Occupy Wall Street protesters: https://theweek.com/articles/480857/demographics-occupy-wall-street-by-numbers.

2   *South China Morning Post*, "Why Hong Kong's angry and disillusioned youth are making their voices heard", 22 July 2019, Thomas Peter.

3   Approx USD$930,000 for a 40sqm (431Sqft) apartment: https://www.reuters.com/article/us-hongkong-economy-property/hong-kong-private-home-prices-rise-at-fastest-pace-in-a-year-in-april-idUSKCN1IW0CD.

4   Global Database of Events, Language, and Tone.

5   "Middle-class rage sparks protest movements in Turkey, Brazil, Bulgaria and beyond", *Washington Post,* Anthony Faiola and Paula Moura; 28 June 2013.

6   See "The Age of Mass Protests: Understanding an Escalating Global Trend", Haig et al, Centre for Strategic International Studies (Risk and Foresight Group), March 2020.

7   "Dole" came from post-WWII benefits that were "doled out" to recipients.

8   "Automation and Artificial Intelligence: How Machines Affect People and Place", Brookings Institute (2019).

9   "Automation and the future of work – understanding the numbers", Oxford Martin School Blog, by Prof Michael Osborne & Dr Carl Frey, 13 April 2018.

10   See; "Solar is now the cheapest electricity in history, confirms IEA", via Carbon Brief, 13 October 2020 https://www.carbonbrief.org/solar-is-now-cheapest-electricity-in-history-confirms-iea.

11   Source: WHO (4.2m), UN (7m), New Scientist (9m).

12   See: *Pittsburgh Post-Gazette*, "Pushing for black lung benefits, coal miners tread a well-worn path to Capitol Hill", by Daniel Moore, 23 July 2019.

13   See: Exxon's *Climate Denial History: A Timeline*, source: Greenpeace.org https://www.greenpeace.org/usa/global-warming/exxon-and-the-oil-industry-knew-about-climate-change/exxons-climate-denial-history-a-timeline/.

14   Source: Centre for Research on Energy and Clean Air (CREA) and Greenpeace Southeast Asia Report, 12 February 2020 – https://phys.org/news/2020-02-air-pollution-trillion-year-ngo.html.

15   Source: *Business Insider*, "Elon Musk doubles down on universal basic income: 'It's going to be necessary'", by Chris Weller, 13 February 2017.

16   Source: *Wall Street Journal*, "Amazon takes over the world", by Scott Galloway, 22 September 2017

17   Source: World Economic Forum 2018.

18 Source: CNN Business, "Mark Zuckerberg supports universal basic income. What is it?", Patrick Gillespie, 26 May 2017.

19 Source: YouTube, "My Chat with Twitter co-founder Jack Dorsey", the SAAD Truth, 5 February 2019.

20 Source: *Business Insider*, "EBay's founder just invested $500,000 in an experiment giving away free money", by Chris Weller, 8 February 2017.

21 Source: YouTube – Is Universal Basic Income a Good Idea?, Singularity University Channel, 11 August 2016.

22 Source: CNBC, "America should give out free cash to fix income inequality", by Catherine Clifford, 2 July 2018.

23 Source: Medium, "Work is more than a source of income", Tim O'Reilly, 28 September 2015.

24 Source: Medium, "The Case for Cash for All", Chris Hughes, 17 May 2016.

25 Source: Quartz, "Y Combinator's president thinks crypto could facilitate universal basic income", by Matthew De Silva, 28 February 2019.

26 See: Twitter status update @AndrewYNg, 9 November 2016, 11:12am.

27 Source: Oxford University, "In pandemic crisis: 71 percent of Europeans support universal basic income", May 2020.

28 See: Valtioneuvosto Stasrädet, Ministry of Social Affairs and Health, "Suomen perustulokokeilun arviointi", Olli et al, May 2020.

29 See Ambri Liquid Metal Batteries as an example: https://ambri.com/technology/.

30 See: SBS News, "Renewables v coal: Blame game over SA blackouts continues", 9 February 2017. https://www.sbs.com.au/news/renewables-v-coal-blame-game-over-sa-blackouts-continues.

31 Source: The Verge, "A Danish company is building a $335 million seawall around New York", Mona Lalwani, October 2014.

32 See: Stella McCartney unveils mycelium-based clothing, https://www.cbc.ca/news/technology/mycelium-fungi-green-materials-1.5954664.

33 For more on work patterns throughout history, go to https://eh.net/encyclopedia/hours-of-work-in-u-s-history/.

34 Rockerfeller University "Working Less and Living Longer: Long-Term Trends in Working Time and Time Budgets".

35 Source: BLS, Overview of the labour force from 1950–2010 and projections to 2050.

36 Source: The U.S. Department of Housing and Urban Development as reported in US News "Housing Unaffordable across the US".

37 Source: Earth.org "Sixth Mass Extinction of Wildlife Accelerating – Study", Jun 2020: https://earth.org/sixth-mass-extinction-of-wildlife-accelerating/.

38 Source. The Harris Group, "Millennials, Fueling the Experience Economy", July 2018.

CHAPTER 8

TECHNOLOGY CHANGES EVERYTHING

"No one pretends that democracy is perfect or all-wise. Indeed, it has been said that democracy is the worst form of government except for all those other forms that have been tried..."
—Winston S Churchill, 11 November 1947

It could be said that Socrates and Plato predicted the rise of "fake news". Socrates argued that, to participate in democracy, a minimum level of education and thoughtful consideration of issues should be required to vote. Otherwise democracy would be easily corrupted.

Democracy in the Digital Age

In Plato's *The Republic* book VI (488a–489d), Socrates gets into an argument with Adeimantus to persuade him that democracy as a form of governance is not effective. He uses a metaphor of the state as a ship to reveal his point. In the metaphor, Socrates compares the population at large to a strong but unsophisticated ship owner whose knowledge of seafaring is limited. The ship's navigator (the philosopher) competently steers the ship, but he is criticized by the sailors as useless because he's always staring at the stars. The sailors (demagogues and politicians) make all sorts of claims in respect to their capability to sail the ship and attempt various schemes to get the ship owner to put them in place as the ship's captain. They ply the ship owner (the people) with drugs and wine to get him on side. Then they dismiss the stargazing navigator, who is the only one with the requisite skills to actually guide the ship through dangerous waters.

For Socrates, voting and participation in a democracy was not a birthright, nor would it be based on citizenship. He argued that unless educational standards were extremely high and citizens engaged thoughtfully on policy, democracy would fail because of the lack of critical thinking skills brought to bear on decisions that affected the state.

Thomas Jefferson (1743–1846) also believed that a strong education system is vital to the effective working of democracy. John Sharp Williams wrote that Jefferson's impact on education is pronounced because Jefferson saw that "democracy and education are interdependent" and therefore with "education being necessary to its [democracy's] success, a successful democracy must provide it"[1]. The fact that comparative education metrics have declined so significantly in the United States over the last few decades along with manipulation of policy and government by vested interests, lends support to Jefferson's view that democracy is weakened by lack of commitment to quality national education.

Putting aside that Socrates and Plato were essentially arguing that philosophers like themselves would make the most competent leaders, it is clear that we need "big picture" leadership across the world to succeed in tackling the emerging crises that are coming at us. It is no accident that as literacy, numeracy, and scientific knowledge have declined in many modern democracies, that fake news and conspiracy theories like the flat earth, "fake" moon landings and such have thrived. Social media and the internet have made facts and nonsense both equally accessible for the general public to consume and have given them a measure of equivalence. Alternative facts, as Trump's press officer Kellyanne Conway labelled them, are not the same as actual facts, but they are both as widespread today. In fact, during the 2016 election, the top 20 fake news stories shared on Facebook outperformed the top 20 legitimate news stories on major mainstream media outlets—8.7 million shares to 7.3 million[2].

Trump argued Mexicans were coming to steal American jobs and they would bring crime with them, neither of which was proven. Trump ran on the promise to bring back coal jobs, which he described as good for the economy, and then declared "Big Coal is back!"[3]. He did so at a time when unsubsidized solar energy has plummeted to one sixth of the cost of coal-based electricity[4], where new renewable jobs outpaced new coal jobs by a factor of 100 times and where coal use shrank by 35% (eliminating 10% of US coal jobs).

Today we don't have the mechanisms to accurately limit the spread of fake news, alternative facts or just plain misinformation. But we're going to need it, particularly to reinforce Artificial Intelligence's impact on curation and contextualization in the future, and to rebuild trust in core institutions such as government and the media.

The problem with digital content is the lack of reputational accountability. In the old days, if you were a journalist or news anchor and you posted fake information, you'd be called out, fired or sidelined. Sources were protected, not only because you wanted to protect the source from a backlash, but because sources brought you informational credibility. Think about Mark Felt, the FBI informant known as "Deep Throat", who helped Bob Woodward and Carl Bernstein expose the Watergate scandal in the Nixon administration. Or Frank Serpico, the NYC police officer who exposed rampant police corruption. Or Jeffrey Wigand, who in 1996 exposed on the *60 Minutes* TV show the fact that tobacco companies had been long aware of the addictive effects of nicotine in cigarettes and had worked to enhance that effect. As a journalist, you couldn't break these major stories without verifiable facts and a credible source, who you protected at all costs.

Today, however, mainstream media has become less about breaking news and investigative journalism and more about ratings. This has led to some of the biggest cable news networks on the planet misreporting, using "click bait", applying their "spin" or even presenting outright falsehoods to garner support from specific audiences. Sources might include conspiracy theorists presenting outlandish positions that are presented side by side with scientists who have dedicated their life to their field, as an example.

The other concern is that technology platforms like Facebook and Twitter have clearly accentuated the dumbing down of citizens and have given rise to the purveyors of fake news and alternative facts. Trump would have been far less effective at getting segments of America angry at Mexicans and China had it not been for social media's ability to amplify information indiscriminately. Unsurprisingly, a study from the University of Baltimore and Israeli cybersecurity firm CHEQ revealed that fake news is already costing the global economy over $78 billion a year.

If we're going to enable fully participating, transparent democracies, then we're going to need to mobilize science effectively against vested interests, misinformation and politics. To achieve that, we're going to need a whole new level of transparency and accuracy. During COVID-19, time and again basic science was undermined by politics. The same is true for climate change, but add in the combined billions of dollars deployed by vested interests to obfuscate the facts and we have a significant headwind against transparency. The internet, while being a boon for commerce and providing access to the sum knowledge of humanity, has also eroded some of the basic concepts we use to regulate the use of facts versus rumour and disinformation in the real world. Social media, while creating incredible vehicles for user growth and engagement, has also created very divisive actions, digital bullying and ostracism, leading to suicides, hate crimes and worse.

It goes without saying that fake news and "alternative facts" get much better mileage when they are promoted by a leading public figure, rather than your average Facebook or Twitter troll. Although QAnon is certainly an example of fake news gone wild, as is the spike in the number of "flat earthers", moon landing and coronavirus deniers, it is also true that today, a farmer in Africa with a smartphone has more information at his fingertips than Bill Clinton did at the end of his presidency.

The web even has a term for how ideas on fringe online subcultures travel to large audiences on mainstream platforms and news outlets—it's called *normiefication*. A great study into the QAnon phenomenon (Zeeuw et al, 2020[5]) showed how Q started as a fringe idea on 4chan around "Pizzagate" (a theory that leading figures in the Democratic Party were involved in a paedophile child-trafficking ring[6]) that then morphed into a major movement promoted by the President himself.

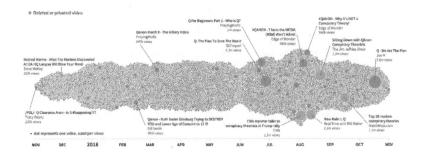

Figure 1: Beehive graph of QAnon videos on YouTube. Data requested on 2 December 2019 with the search endpoint of the YouTube v3 API. Sized per video views and coloured on availability on 22 October 2020. (Credit: Zeeuw et al, 2020[7])

What's the solution to all this false data?

Clearly improving access to cheap, high-quality education through technology is vital for the advancement of the species as a whole. For most of the 20th century aggregate IQ scores were improving, adding about three IQ points per decade (see the Flynn effect), but since 1975 we've seen that trend reversing[8]. We need to address that trend, and we'll also need to augment our intelligence using technology—although as Elon Musk argues, we're already doing that with full-time internet access via the devices around us.

Better high-quality affordable education will certainly allow us to combat fake news, but we also need greater transparency and greater accountability. If you post verifiably false information or propaganda repeatedly, then you should simply lose the privilege of influence. We also need to be transparently clear about where the information has originated.

Never has transparency been as possible as it is today. But this transparency requires a new framing of information and data sources according to some core ethical construct that measures truthfulness, accuracy, and factualness. Those ethics must be separate from politics and commercial interests—they must be honestly optimized for humanity as a whole. We'll also need the same transparency to ensure unbiased and equitable Artificial Intelligence (AI).

Organizing principles for optimal human advancement

Would a benign autocracy of the kind Plato suggested work better for human advancement? It is possible. Singapore, for example, operates in roughly the manner that Plato outlined and does very well economically. Singapore ranks highly in education, with strong science, math and literacy scores. Singapore has had one of the best-performing economies in Asia over the last 50 years, too. Interestingly, research indicates that Singapore achieving the highest level of economic development in Asia over that period (along with a higher level of per capita GDP than the US) is based mostly on massive accumulation of capital and then subsequent investment in the labour force. Whereas productivity gains, which is favoured as the measure for economic improvement in much of the West, has played a smaller role in Singapore's economic success.

It's not all sunshine and roses, however. Sixty percent of Singaporeans are generally happy with their government, but in World Happiness rankings they sit below countries like America and Australia. Keep in mind that only 17–18% of Americans report as content with their government.

In the World Happiness Report (an annual report supported by Gallup, Oxford University, Columbia's Centre for Sustainable Development, and others), Nordic countries—which include Denmark, Norway, Sweden, Finland, and Iceland—consistently score as the happiest nations on earth. And that's despite long winters, big taxes and Abba.

What makes citizens happy then? Is freedom the basis of happiness?

Freedom House publishes an annual "Freedom in the World"[9] report that documents the changes in freedoms by country. For 2020, 141 countries ranked as "free" or "partly free", and 54 ranked as "not free". New Zealand, which has performed admirably during the coronavirus pandemic, was one of the highest-ranked countries in terms of freedom in 2020 (ranking 99 on the freedom scale). Finland, Norway and Sweden all scored 100, and the United States 83. The US, a country that prides itself on its freedom, doesn't even rank in the top 20 countries in respect to "freedom".

Unfortunately, globally "freedom" is on the decline and the 2020 coronavirus pandemic was a trigger for the largest measurable decline in net freedoms that we've seen in the last 15 years. This spells trouble for the world of the 21st century, as rolling crises will enable political gaming like never before and therefore, fewer freedoms if the pandemic is any indication.

Here's the thing. The pandemic allowed for the abuse of political powers not because of the coronavirus itself, nor because of scientific priorities. It was simply opportunistic politics at worst, and administrative incompetence at best.

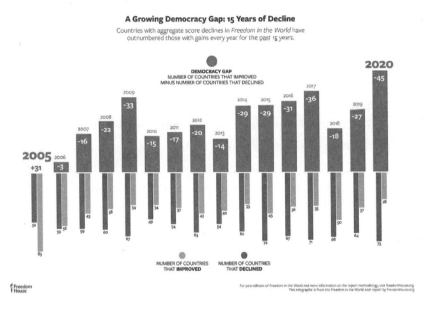

Figure 2: Freedom has been on the decline for the last 15 years.
(Source: Freedom House)

Beijing's stated non-interference doctrine (in what they articulate as internal politics) and the application of the Article 23 anti-sedition law in Hong Kong, shifted Hong Kong from one of the freest states to partly free. The United States saw a significant decline over the last decade, moving from 93 to 83 on the Freedom House ranking, driven by political corruption, lack of transparency in government and more recently punitive immigration and asylum policies. In Hungary, Prime Minister Viktor Orbán exploited the pandemic to amass emergency powers to challenge the press and eliminate criticism of his administration. In the Philippines, President Duterte banned "false information" that criticized his administration's response to the coronavirus crisis.

Why is this important? Freedom correlates quite strongly with citizen happiness. Twelve of the top 20 freest countries are also in the top 20 happiest countries to live.

And how does economic performance correlate with happiness? Macroeconomics appears almost completely unconnected to both freedom and happiness overall. Only two countries appear in all three top 20 indexes (freedom, GDP per capita and happiness rankings). And those two countries are? Norway and Luxembourg. Incidentally, the two largest economies in the world, the US and China, don't rank in the top 10 countries in terms of freedom, economics or happiness.

	FREEDOM		ECONOMICS		HAPPINESS	
Rank	Country	Score	Country	GDP Per Capita	Country	Score
1	Finland	100	Liechtenstein	139,100	Finland	7.809
2	Norway	100	Qatar	124,100	Denmark	7.646
3	Sweden	100	Macau	122,000	Switzerland	7.560
4	New Zealand	99	Monaco	115,700	Iceland	7.504
5	Canada	98	Luxembourg	105,100	Norway	7.488
6	Netherlands	98	Bermuda	99,400	Netherlands	7.449
7	Uruguay	98	Singapore	94,100	Sweden	7.353
8	Australia	97	Isle of Man	84,600	New Zealand	7.300
9	Denmark	97	Brunei	78,900	Austria	7.294
10	Ireland	97	Ireland	73,200	Luxembourg	7.238
11	Luxembourg	97	Norway	72,100	Canada	7.232
12	Belgium	96	Falkland Islands (Islas Malvinas)	70,800	Australia	7.223
13	Japan	96	United Arab Emirates	68,600	United Kingdom	7.165
14	Portugal	96	Kuwait	65,800	Israel	7.129
15	Switzerland	96	Hong Kong	64,500	Costa Rica	7.121
16	Barbados	95	Switzerland	62,100	Ireland	7.094
17	Slovenia	95	Gibraltar	61,700	Germany	7.076
18	Cyprus	94	United States	59,800	United States	6.940
19	Estonia	94	San Marino	59,000	Czech Republic	6.911
20	Germany	94	Jersey	56,600	Belgium	6.864

Figure 3: The top 20 countries as ranked by freedom, GDP per capita and happiness. (Author's own)

Economic performance would appear to rank significantly lower in importance to the average citizen than political and social freedoms, good governance, good health, longevity and social support systems. Even then, Chinese citizens, for example, arguably experience fewer individual freedoms than in Western democracies, but they're also generally happier with the government and feel they have much greater freedoms today than in the past—largely afforded by improvements in economic conditions. The Harvard Gazette published an Ash Centre survey-based study that looked at the level of satisfaction with the Chinese central government from 2003 to 2016. They found across China there were high levels of satisfaction with the central government with 95.5% of respondents replying they were "relatively satisfied" or "highly satisfied" with Beijing.

"Every time the World Happiness Report has published its annual ranking of countries, the five Nordic countries – Finland, Denmark, Norway, Sweden, and Iceland – have all been in the top ten... No matter whether we look at the state of democracy and political rights, lack of corruption, trust between citizens, felt safety, social cohesion, gender equality, equal distribution of incomes, Human Development Index, or many other global comparisons, one tends to find the Nordic countries in the top spots."
—"The Nordic Exceptionalism: What Explains Why the Nordic Countries Are Constantly Among the Happiest in the World", World Happiness Report, 2020

Looking at the most successful economies with happy citizens, you just don't get the picture that the media's portrayal of what makes a country attractive actually aligns with what we see in terms of research and feedback. Those living in the Nordics might occasionally complain about the high taxes, but they generally understand that the tax system gives them access to a quality of life that simply is not achievable elsewhere. This is similar for those living in China. They're immensely proud of the progress China has made as a world-leading economic force, but more so because the average Chinese can see that the government has been working hard to improve the lives and affluence of the middle class.

Prioritizing happiness

The leading economies of the future will not only deliver a standard of living for people to be happy, but will attempt to give citizens greater freedoms independent of their job or net worth. This will require that people be provided with access to a quality of life, a range of experiences and opportunities for self-development that fulfil them.

To this end, the United Nations Development Program's Human Development Index (HDI) attempts to capture features related to standard of living that go beyond pure economics. It combines assessments of longevity, health, education, "being knowledgeable" and having a decent standard of living, to arrive at an overall view on which places presently offer the best standard of human development.

Rank	Country	HDI value (2019)	Life expectancy at birth (years)	Expected years of schooling (years)	Mean years of schooling (years)	Gross national income (GNI) per capita (PPP $)
1	Norway	0.957	82.4	18.1	12.9	66494
2	Ireland	0.955	82.3	18.7	12.7	68371
2	Switzerland	0.955	83.8	16.3	13.4	69394
4	Hong Kong, China (SAR)	0.949	84.9	16.9	12.3	62985
4	Iceland	0.949	83	19.1	12.8	54682
6	Germany	0.947	81.3	17	14.2	55314
7	Sweden	0.945	82.8	19.5	12.5	54508
8	Australia	0.944	83.4	22	12.7	48085
8	Netherlands	0.944	82.3	18.5	12.4	57707
10	Denmark	0.94	80.9	18.9	12.6	58662

Figure 4: United Nations Development Program's Human Development Index (HDI).

Norway ranked first overall in the HDI[10] in 2019, Ireland and Switzerland ranked equal second, Hong Kong and Iceland ranked equal fourth, Germany sixth, Sweden seventh, Australia and Netherlands equal eighth and Denmark tenth.[11]

Both Hong Kong and Switzerland are in the top 10 for both the HDI and Global Knowledge Index (GKI) 2020[12], the most recent Fraser Institute Economic Freedom of the World Rankings and the most recent HDI. Hong Kong and Switzerland are countries for others to emulate in the present knowledge economy.

The countries that are performing well in global rankings for human development, knowledge, and economic freedom are likely delivering a quality of life that will make their existing workforce relatively happy and which could be used to attract mobile professionals from around the world. The more interesting, intelligent and creative people that a place can attract, the more likely that a snowball effect will occur, drawing more and more talented people to the location and making it more likely to continue to succeed.

What would an optimal economy built for its citizens look like? The first criteria is that it would be prioritized around satisfying the basic needs of citizens. In 19th century terms we might be tempted to call this model communist or socialist, and it does tend towards social prioritization; but in pure economic terms it essentially looks like Durant's diamond-shape model, where economic stimulus is designed to create employment, consumption, and wage growth across the board, versus just wealth creation. A diamond-shaped economy can still be overtly capitalist, as we saw with the United States' economy in the 1950s–1970s. But the US post-World War II prioritized getting the GI servicemen back to work, advancing infrastructure to provide a competitive foundation, and getting the economy moving for the middle-class after the austerity measures taken during the war.

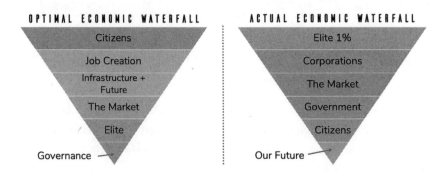

Figure 5: How the economy should work for citizens vs how it works today (author's own)

We can make a fairly solid argument that this optimal economic waterfall model is overall better performing than one that is prioritized around the wealthy and corporations. Especially considering that the Laffer curve (tax cuts on the rich and corporations designed to flow down to the middle class) has consistently failed to bolster middle-class growth in the US—it's a simple matter of deduction that wealth accumulation at the top of the market doesn't necessarily create GDP growth compared to simple wage growth across the economy.

Another essential component is investing for a future state. Clearly the global economy is coming together in a way that was never possible when Adam Smith or John Maynard Keynes wrote their economic theses, and while Friedman lived to see the birth of the commercial internet, it certainly didn't factor into his most well-known work, *Capitalism and Freedom* (1957). Friedman, of course, is a favoured economist for conservatives because of his position on minimal government interference and his stance against the welfare state. Friedman's answer to poverty and inequality was that richer, well-off people should all agree to contribute something, so they didn't have to see poor people in their neighbourhoods anymore. That's a simplification, but his broad argument was that the economy first and foremost should generate income and governments could worry about the social issues later.

But coming back to the internet. Clearly this is a world where global commerce is the norm, not the exception, where an American can order something on Amazon, and an Amazon reseller can then fulfil that order via Alibaba by air-freighting from China to the US just 24 hours later. This is a globally interconnected economy, but one that clearly relies on significant technology investment and infrastructure to thrive. An economy based on renewable energy is not only going to be cleaner and cheaper, but is increasingly future proof, while creating strong new job growth. An economy that invests in community health through better policy, purposeful data management such as genome sequencing of the population and the use of AI, will create healthier, longer living citizens. These investments are the opposite of Friedman's model, because it is only when the existing system stops making its highest returns that the market would shift core working practices and investment in industry.

With technology, AI and smart infrastructure being so fundamentally critical for 21st century competitiveness, the need to prioritize investment in said systems and support an aggressive retooling of the economy and individual skill sets, requires a much greater investment up front than Friedman's form of capitalism allows for. This may explain why the US economy is rapidly falling behind China's. It's not only consumption that matters, but also economic incentives.

Those economies where there is plenty of debate about freedom, inequality and democracy, are the places that still struggle to provide an equal playing field, or where the conditions for the average person are declining. China has fewer of these debates because the average constituent is simply happier. Finland, Norway, Denmark and Sweden also have fewer of those debates because citizens know their needs are prioritized over that of corporations and the wealthy. In the US, the tired old debate that if homeless people really wanted a roof over their head they'd work harder, or that the only reason someone's health costs made them bankrupt is that they should have educated themselves better so they could get a better job, is based on a flawed premise. It's based on the fact that most Americans work for an economy designed to enrich a small percentage of the population, not an economy that prioritizes improving access to core services as it grows.

Winners versus losers: China's 21st century advantage

Let's face it: in this day and age with the resources we have access to, no citizen should go without the opportunity to better themselves, or to provide a healthy living standard for themselves and their family.

The world total global wealth was estimated at $360.6 trillion in 2019. While that's only approximately $50,000 per person, the value that the economy adds to each individual is separate and distinct from that.

We come back to that fundamental question we asked in both Chapters 1 and 4: what is the real purpose of the economy? Is it to create economic growth, or to serve people's needs? If you want happy, fulfilled citizens, it's the latter. If you want a successful 21st century economy, ironically it may also be the latter. What supports this view?

As discussed earlier, the period from the end of World War II to the early 1970s was one of the greatest eras of economic expansion in world history. US GDP grew from $228 billion (1945) to around $1.7 trillion (1975) in just 30 years. By 1975, the US economy accounted for 35% of the entire world's industrial output and was over three times larger than that of Japan (the world's second-largest economy). But the real secret sauce was middle-class growth. The economic growth experienced in the US was distributed fairly evenly across the economic spectrum—this stimulated consumption, housing, manufacturing, the auto industry, and the electronics sector. Much of the growth came from the movement of low-income farm workers into better-paying jobs in the towns and cities. America put men on the moon, had incredible advances in medical science and in technology advancement. The entire populace was inspired by the possibilities, and they knew that their children would be better off.

In the 1950s, 90% of the world's middle class resided in Europe or the United States with a negligible percentage in China. But today, roughly 20% of the global middle-class live in China. By 2027, 1.2 billion Chinese will be classified as middle class, making up at least 25% of the global total[13].

China already has the largest single consumer market, which means it is starting to look a great deal like the post-World War II US economy in terms of broad, grass-roots growth. Having said that, China has significant advantages over the US economy of 70 years ago, and they are clearly investing at much higher levels in 21st century infrastructure and skill sets that will keep their economy relevant. China is morphing from a classic pyramid-shaped economy (throughout the mid-20th century) to a diamond-shaped economy that characterizes the most profitable economies we've examined historically.

Let's do a quick comparison of the US, China, Japan, India, and the European Union (EU), examining infrastructure development, AI, R&D, core skills development and middle-class growth.

Infrastructure

When it comes to infrastructure investment, China is spending almost twice that of Japan and India and more than three times that of the US and the EU. President Xi Jinping proposed the Belt and Road Initiative (BRI) in a pair of speeches back in 2013. Today, China has invested around $3.5 trillion across 2,881 projects in 70 countries spanning the globe[14]. By the completion of the Belt and Road Initiative (BRI) in 2050, it is estimated China will have spent something in the order of $8 trillion. It has been billed as the largest and most ambitious infrastructure mega-project of the modern era. Once completed, it will stretch from East Asia to East Africa and Central Europe.

The project includes the obvious trade route developments such as rail networks, road networks and maritime shipping capacity and investment. The BRI initiative will transform the Eurasian landmass, reshape global trade and lead to strong competition from the renminbi against US dollar-based trade. The BRI infrastructure will touch 70 countries and impact more than 60% of the world's population and 40% of global economic output.

In comparison, Joe Biden's new infrastructure plan proposed approximately $2 trillion of spending to refresh American infrastructure.

"[Biden's] far-reaching American Jobs Plan includes spending to repair aging roads and bridges, jump-start transit projects and rebuild school buildings and hospitals. It would also expand electric vehicles, replace all lead pipes and overhaul the nation's water systems... build the nation's clean energy workforce, expand manufacturing and boost caregiving as a profession to serve the elderly and disabled."
—*USA Today*, "Joe Biden wants to spend $2 trillion on infrastructure and jobs", by Javier Zarracina, Joey Garrison and George Petras, 2 April 2021

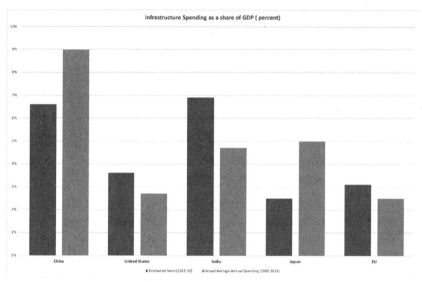

Figure 6: Global infrastructure need/spend as a percentage of GDP.
(Source: author's own)

Most of the infrastructure funding for massive deep-water ports, high-speed railroads, and green-energy power plants will be coming from the coffers of Chinese state-owned companies. Some will be grants, but many are taking the form of loans, and when countries default there can be significant consequences. Between 2006 and 2017, Kenya took out infrastructure loans from China of around $9.8 billion (Sh1043.77 billion). Today, Chinese debt accounts for approximately one fifth of Kenya's foreign debt, and more than two thirds of Kenya's bilateral trade debt. The funds were used for projects like highways, roads, and a railway that stretched between Mombasa and Nairobi.

In late December 2018, Kenya reportedly came close to defaulting on Chinese loans to develop its largest and most lucrative port in Mombasa. A default could have forced Kenya to relinquish control of the port to China[15]. Similarly, in Pakistan, a deep-water port in Gwadar was funded by loans from Chinese banks to the tune of around $16 billion. The problem is that with an interest rate of over 13%, it means the risk of default is significant. In the event of a default, China could end up claiming all sorts of collateral as compensation. Collateral already held by Chinese players includes everything from coal mines, airports and freeways to oil pipelines and railways.

But China isn't likely to call in those loans except in exceptional circumstances. China doesn't want to own infrastructure in a sort of neo-economic colonialism, they want to dominate global trade. The incentive to drive Chinese participation in the global economy is frankly much stronger in China than the rest of the world, and much stronger than the opportunity to take control of state-owned assets and infrastructure at a local level. For sure, there will be some defaults, but China understands that the more defaults they leverage off, the less likely the Belt and Road Initiative will work to encourage trade with China over the longer-term, as developing countries would get skittish if they see China taking over core economic assets.

China also understands that trade and development are tied very strongly to the deployment of AI capabilities. In fact, we could look at AI as part of the infrastructure development that China is going through, not only in R&D spend and startup investments, but even in the way they are teaching primary through university age students about AI and its potential impact on their lives.

Artificial Intelligence and R&D spend

In 2017, the State Council of the People's Republic of China published the Artificial Intelligence Development Plan[16]. This is core to China's future economic plans, as well as forming an integral part of the digital Silk Road initiative that parallels their BRI investments. The document makes very clear China's stated objective: to lead the world in the use, development and application of Artificial Intelligence by 2030 or earlier.

"The rapid development of artificial intelligence will profoundly change human social life and the world. To seize the major strategic opportunities for the development of artificial intelligence, build China's first-mover advantage in artificial intelligence development, accelerate the construction of innovative countries and the

world's science and technology power, this plan is enacted
in accordance with the requirements of the CPC Central
Committee and State Council."
—Notice of the State Council Issuing the New Generation of
Artificial Intelligence Development Plan, 8 July 2017

Some studies show that the US has spent slightly more on Artificial Intelligence than China over the last 5–10 years, but those studies focused on US venture capital investments in AI startups and their investment in development of AI for military use, etc—these studies don't take into account wider investments in AI across society (where China arguably leads). The US Department of Defense has repeatedly raised concerns the US is falling behind significantly in AI investment. The US military estimated that China's spending on AI grew to at least $70 billion in 2020. In contrast, the Pentagon had plans to invest about $4 billion in AI and machine-learning research and development in 2020.

The Economist raised concerns back in July 2017 that China was already overtaking the US in critical application of deep learning capabilities[17], calling it the "Saudi Arabia of Data" (if data is the new oil). China has 730 million internet users, more than twice the 312 million users in the United States. Thus, when it comes to utilization of deep-learning or AI capabilities, China consistently shows much faster take up of AI-based technologies. Consider, for example, mobile wallets. Apple expects to be approaching $1 trillion of mobile wallet spend by 2025, whereas China surpassed that way back in 2013, putting the US potentially 12 years behind China in terms of mobile wallet adoption. In 2020, estimates range between $52–58 trillion of mobile payments made across Alipay and Tencent WeChat Pay alone. In contrast, global plastic card payments (including credit cards, debit cards, prepaid cards and gift cards) topped out at $25 trillion in 2017, and the most ambitious estimates have total card spending at $45 trillion by 2023. If you're running the math, that means China transacts in mobile payments almost twice what the entire world does in plastic card payments.

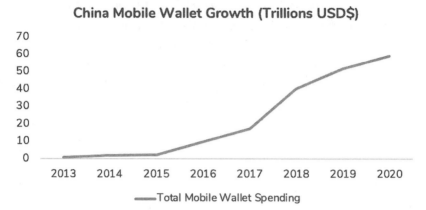

Figure 7: China's mobile wallets incorporate facial recognition at a national scale.
(Source: author's own)

China's massive growth in mobile payments is clearly dependent on their internet connectivity, but mobile wallet growth also demonstrates the smart use of facial recognition technology, and AI used in tracking down fraud and identity theft. On 11 November 2020, Alibaba broke all its previous records by racking up $56 billion in sales in a single day, the vast majority of these transactions going across Alipay's mobile payment network, utilizing biometric security. The use of this form of AI technology means that even though Alipay was processing at peak 459,000 transactions per second, they only recorded 0.0006 basis points of fraud (bps)[18].

Figure 8: Jack Ma demonstrating facial recognition payments technology back in 2014.
(Credit: Ant Group/Alipay)

In comparison, US credit cards have received on average of 2.92 bps for "card present" transactions and 11.44 bps for "card not present" transactions (which are typically online e-commerce payments)[19] during the last decade. On that basis, facial recognition technology is an order of magnitude safer than chip and pin, or signature-based transactions on plastic cards. This means we must move away from cash and plastic over the coming decade with unsustainable fraud rates and crime. While the US and EU has thus far publicly rejected facial recognition technology (allegedly for its civil rights issues), the reality is they both use it widely in respect to law enforcement and for driver's licenses, passports and border control. This is probably not understood by the average citizen.

By 2030, national biometric-based identity schemes will be standard across the world where we are using the internet to access day-to-day services, as will passports based on technology like facial recognition and the blockchain. By the time the West gets there, China will already have had a decade-and-a-half of experience in using this tech.

China's ambitions in respect to artificial intelligence and to research and development (R&D) generally are clear. In 1999, the World Intellectual Property Organization (WIPO) received just 276 patent applications from China; in 2019 that had exploded to 58,990 patents[20]. The US recorded 57,840 patents in that same year, with Japan coming in third at 52,660 patents. Huawei filed 4,411 of these patents alone. China's AI investments and filings show prioritization around the following AI-based categories:

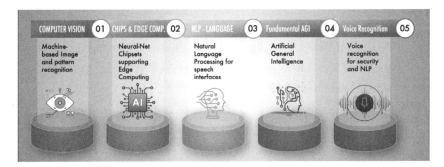

In August of 2020, China also surpassed the US in respect to scientific papers filed. China accounted for 19.9% of scientific papers, while the

US came in second at 18.3%[21]. China remains number two globally for research and development spending for now, but is expected to overtake US R&D spend by 2022.

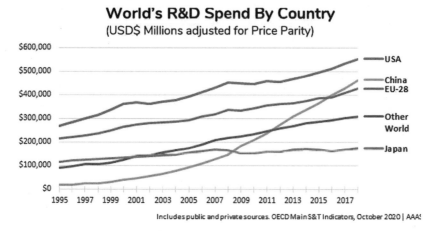

Figure 9: Global spend on R&D last two decades. (Source: OECD figures)

Growing technology-based skills gap

Where China really excels globally is its commitment to developing a future-proof citizenry. While India (78 million) tops both China (77.7 million) and the US (67.4 million) for total number of university students, by 2016 China was building the equivalent of one new university every week. This is addressing the gap in educating students for a technology-led industrial and services economy.

The World Economic Forum reported that China had 4.7 million new STEM graduates in 2016, while India had 2.6 million and the United States had just 568,000 STEM graduates in the same period. It is estimated that China will have nearly 40 million STEM graduates annually from 2020. While some might argue China's science and engineering programs aren't as sophisticated or developed as those in the US, the fact remains that in a typical year 40–50% of Chinese bachelor's degree graduates are those studying in STEM disciplines, whereas in the US it is in single digit percentages. This means that by 2030, China's STEM workforce will have grown 300%, versus the US' 30% growth—and even then, that incorporates the fact that the US still relies heavily on H1-B visa immigrants to fill out the totals (although typically accounting for just 0.5–0.7% of the US college graduate labour force).

For economies dramatically geared towards technical dominance, retooling manufacturing capabilities towards autonomous operation, mass-scale system design using deep learning of national-wide data sets, and retooling energy systems towards smart distributed energy storage and renewables, the sheer amount of STEM talent a country is capable of producing is easily the number one concern for a future-proof labour force. In the US, the reliance on H1-B at the high end of the STEM community is very clear. Renowned US theoretical physicist, Dr. Michio Kaku, famously called the H1-B program the US secret economic weapon. Here is why.

More than a third of US' doctoral chemists are non-US citizens or naturalized foreigners; 53% of US doctoral candidate engineers were non-US citizens; 76% of foreign-born recipients of US science and engineering doctoral degrees hope to stay in the US; and 24% of US patents have at least one US non-citizen listed as an inventor. College educated, foreign-born scientists and engineers comprised almost 30% of the entire science and engineering workforce in the US as of 2017[22].

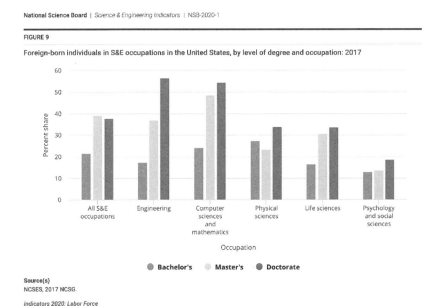

National Science Board | *Science & Engineering Indicators* | NSB-2020-1

FIGURE 9

Foreign-born individuals in S&E occupations in the United States, by level of degree and occupation: 2017

Source(s)
NCSES, 2017 NCSG.

Indicators 2020: Labor Force

Figure 10: Percentage of foreign-born scientists and engineers in the US by occupation. (Source: NSF)

As noted earlier, the US generally performs poorly in math and science education. In the most recent Programme for International Student Assessment (PISA) results from 2018, the United States ranked 36th out of 79 countries in math, 24th in science, and 13th in reading. Among the 35 members of the OECD (PISA's principal sponsor), the United States ranks fifth from bottom in math. The bigger problem is that PISA's standardized scoring system shows US students vary wildly in their performance, depending on their economic status.

In the US, 93 points separate the average score in the poorest schools from the wealthiest. That's about three grade levels, or the difference between a typical 10th grade achievement and 7th grade achievement. Some 27% of advantaged students in the United States, but only 4% of disadvantaged students (compared to OECD averages 17% and 3%, respectively), were top performers in reading (having attained one of the two highest proficiency levels).

When it comes to future-proofing the economy through STEM skills, the US has a problem. Not only has the relative quality of US education in this area not improved PISA performance in the last 30 years, but the reliance on foreign-born STEM workers in the labour force has significantly increased. While the H1-B visa program has allowed the US' largest tech companies to thrive, pressure on this form of immigration is acute today, meaning the US must rely on homegrown skills. The only way for the US to rapidly turn this around is to provide broad access to STEM programs for free across the nation, and stimulate job programs from top tech employers, which is barely happening today.

China, however, has very broadly committed to what they call "core competencies" or "*hexin suyang*" around STEM and AI. China's educational goal has been established under a fundamental realization of the relationship between AI and human intelligence, collaboration between humans and machines, and how AI is tied very strongly to future developments. Education in China is transforming from a traditional systemic emphasis on the completeness of academic knowledge to much more focus on skills training focused on raising students' abilities and the quality of their thinking and problem-solving abilities[23]. Chinese high school IT curriculum, for example, no longer simply centres on computers and the internet (typically called ICT in US, Australian and UK schooling systems), instead focusing on data, algorithms, information systems and the information society. China's adaptation of its schooling curriculum, its aggressive spending on R&D, patents, AI deployment and technology integration belies the stark ignorance behind those that might claim China is just copying or mimicking US intellectual property[24].

In any case, if you want your economy to be relevant for the 21st century, follow China's lead. Their focus on the fundamental skills, infrastructure and underlying technology development for the next decade and beyond shows a commitment unseen elsewhere in the world. This is why China will not only be the world's largest economy by 2030, but the core producer of the most significant technologies used globally throughout the remainder of the century. Maybe the first trillionaire will be a Chinese AI creator after all.

Developing countries no more

The four-leading so-called "developing" countries are Brazil, Russia, India and China—or the BRIC economies, as we like to acronymize them. But by 2050 these so-called "developing" economies will have outpaced all but the US when it comes to the top GDP performers globally.

Although the United States will still be the third-largest economy in 2050, it will be much smaller than that of China, which will probably be 1.5–2 times the size of the US economy by this stage. Goldman Sachs projects that China's GDP should match America's by 2027, and then steadily pull ahead. India will have met or surpassed the US by 2030 also.

The collective GDP of the four BRICs is likely to match that of today's leading Western nations by 2032. The World Bank predicts that the US dollar will lose its global dominance by 2025 as the dollar, euro and China's renminbi become co-equals in a "multi-currency" monetary system.

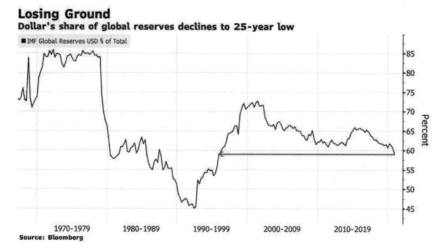

Losing Ground
Dollar's share of global reserves declines to 25-year low

Figure 11: US dollar's share of global reserves is at its lowest level since 1995 .
(Credit: Bloomberg)

It is clear that in the 21st century, what constitutes a "developed" versus "developing" nation needs to be reassessed. The United Nations Department of Economic and Social Affairs created the World Economic Situation and Prospects (WESP) classification system for economic development based on economic research that started in 1975. Today, the G7 incorporates the major developed economies of Canada, Japan, Germany, France, Italy, United Kingdom and the United States. By 2018 other WESP-listed advanced economies included Australia, South Korea, Netherlands, Spain, Switzerland, and Taiwan. The G20 rounds out the biggest economies with what WESP still classifies as developing nations, including Russia, China, Brazil, Argentina, Indonesia, Mexico, Saudi Arabia, South Africa and Turkey.

The G20 is the international forum that brings together the world's major economies annually to discuss macro-economic policy formation and cooperation. Today, the G20 members account for more than 80% of global GDP, 75% of global trade and 60% of the world's population.

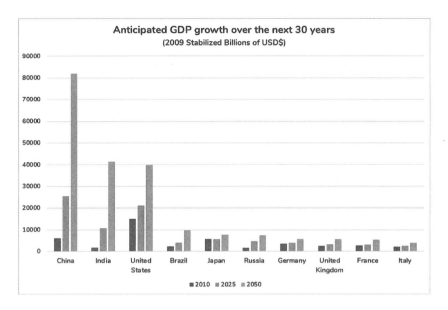

Figure 12: GDP growth and ranking 2050. (Sources: PwC, WEF, OECD)

But taking a long-term view over the next 30 years, what could change? Which economies could become supercharged like the post-World War II US economy?

Clearly China is in the lead across multiple key metrics. Not only is the nation already on track to be the dominant global economy, but China has also invested heavily in global infrastructure that supports global trade and economic dominance well into the future. The Belt and Road Initiative gives China an almost unbeatable platform to develop enhanced supply-chain management and strong economic levers with economic partners, including large-scale infrastructure development loans and investments, as well as favoured trading status across a wide range of cross-border trade.

China's understanding of how AI will define their economy and industries over the next two to three decades appears to be much stronger than the US and Europe. This means that their economic lead for the rest of this century may be unassailable.

Conclusions

It is our estimate that by 2050 AI will account for 25% of global GDP. The greatest economic gains per nation from AI will be in China (26% boost to GDP by 2030) and North America (14.5% boost), equivalent to $10.7 trillion, with these two countries accounting for almost 70% of the global economic impact from Artificial Intelligence. Overall, the biggest absolute sector gains will be in retail, financial services, and healthcare as AI increases productivity, product value and consumption.

Climate will account for at least $7.9 trillion of impact annually in 2050, and that's just in climate-related weather incidents, such as flooding. It doesn't take into account farming impact, mass migration numbers, climate mitigation as a growing industry, and so forth. A loss of 3–7% of global GDP annually is a fairly conservative estimate for climate change negative effects on economic growth. By 2050 every major policy in government will be shaped by climate considerations. Gen-Y and Gen-Z will be firmly in charge of policy and will have accelerated global cooperation, commitment and spending in response to climate disasters globally. It is likely that 20% of the global economy will already be dedicated to climate-related response and mitigation by this stage.

21ST CENTURY ECONOMIC BUILDING BLOCKS

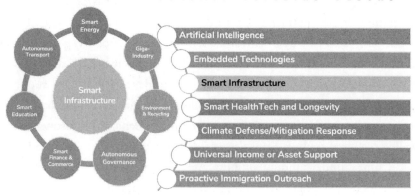

Figure 13: The quintessential 21st century "Smart Economy" requires a complete overhaul of competitive thinking and investment. (Source: author's own)

In fact, it is clear that within the next decade, right next to GDP performance will be carbon neutrality, green energy scores and transparent sustainability commitments. There will be massive international pressure on nations to become carbon neutral and to contribute proactively to climate mitigation programs, both across borders and at home. Sustainable production, recycling technology development, carbon-neutral smart energy management, removal of plastics from water sources, and improvements in air and water quality, will all be areas where nations will be measured and held accountable by the global community.

Here's the key point: the foundation for successful economies is clearly shifting. Tech, science, and engineering skills and investment will be core elements of economies that compete in the 21st century. This requires adaptation to the new normal, and as we've seen through the coronavirus pandemic, some are extremely resistant to this level of adaptation. The politically conservative view emerging in parts of the US, UK and Australia, that we need to return to simpler times (driven by a nostalgia for the 1950s and 1960s), is a psychology that makes technology adaptation increasingly problematic to sell to the citizenry as a whole. Without adaptation, though, you don't get economies that live with AI and climate change in an orderly fashion. You get chaotic resistance that weakens the economy over the longer-term, as policy conflicts with core economic drivers. As Thomas (not Milton) Friedman put it:

> *"It's not the strongest that survive, it's not the smartest that survive, it's the most adaptive that survive...*
> *We are in the middle of a giant adaptation challenge...*
> *at the individual level, at the community level,*
> *and at the corporate level."*
> **—Thomas Friedman, Fortune's Brainstorm Tech conference,**
> **19 July 2018**

Thus, we're getting a clear picture of what the global economic outlook in 2050 will be like. Massive injection of science and engineering skills will continue throughout that entire period; at no point in the next century will the labour force grow against technology-biased skill sets, except where it intersects with climate resilience and mitigation programs. Climate mitigation will be a huge boon for creating new labour demand, but again the bulk of these skills will still be reliant on STEM education. Economies will shift their posture from one of resource utilization to resource maximization—using the resources we have in sustainable and positive ways.

GDP growth, while not being the only measure of success for economies, will be very dependent on core STEM capabilities. China knows this. The US, UK and EU might also know this, but their education systems and investments in their citizens does not currently reflect this knowledge.

The economies of the 21st century will serve their citizens first and foremost as our children and their children become actively involved in setting policy, unconvinced by the economics of the 20th century. The baseline requirements will be that an economy looks after its population first, the environment second, and corporations, the military-industrial complex and politicians a distant last. Unless corporations have an active plan to replace workers displaced by automation, they will face boycotts and huge brand and reputational issues. The same will apply to their climate response.

Think of it like China when the Great Wall was being built, the USA when they were in a race to put men on the Moon, or the world during World War II. At those times the economy focused on something other than just economic growth, and incredible advances were made rapidly. Of course, the purpose of the global economy must be the survival and longevity of the species.

Expect that, from the 2030s through to the later part of the century, humankind will have a collective epiphany that working to protect the planet, to reinvigorate the environment, to save or even resurrect extinct species, to bring some sort of balance between humanity and the natural world, won't be just a nice thing to have—it will become a core philosophy of humanity borne out of necessity because of the destruction we have already wrought on the planet.

It's not adapt or die. It's adapt or go extinct as a species. That's hardly a choice.

Endnotes

1   *Thomas Jefferson: His Permanent Influence on American Institutions*, John Sharp Williams, 1967 (p. 266, 286).

2   Source: Vox.com, "The top 20 fake news stories outperformed real news at the end of the 2016 campaign", Timothy B Lee, 16 November 2016: https://www.vox.com/new-money/2016/11/16/13659840/facebook-fake-news-chart.

3   See *New York Times* - https://www.nytimes.com/2020/10/05/us/politics/trump-coal-industry.html.

4   Source: Lazard's Levelized Cost of Energy Analysis (https://www.lazard.com/media/451419/lazards-levelized-cost-of-energy-version-140.pdf).

5   First Monday, "Tracing normiefication: A cross-platform analysis of the QAnon conspiracy theory" by Daniël de Zeeuw, Sal Hagen, Stijn Peeters, and Emilija Jokubauskaitė, Volume 25, Number 11, 2 November 2020: https://firstmonday.org/ojs/index.php/fm/article/download/10643/9998.

6   Marc Fisher, John Woodrow Cox, and Peter Hermann, 2016. "Pizzagate: From rumor, to hashtag, to gunfire in D.C.," *Washington Post*, at https://www.washingtonpost.com/local/pizzagate-from-rumor-to-hashtag-to-gunfire-in-dc/2016/12/06/4c7def50-bbd4-11e6-94ac-3d324840106c_story.html.

7   Credit: Zeeuw et al, visualization by RAWGraphs: https://oilab.eu/normiefication/#beehive-youtube.

8   Source: Science Alert, "IQ Scores Are Falling in 'Worrying' Reversal of 20th Century Intelligence Boom", Peter Dockrill, 13 June 2018.

9   Source: Freedom House, "Freedom in the World 2021": https://freedomhouse.org/countries/freedom-world/scores?sort=asc&order=Country.

10 United Nations Development Program, *Human Development Report 2020*, New York, 2020.

11 *Economic Freedom of the World: 2020 Annual Report*, The Fraser Institute.

12 See Chapter 9 for more detail on the GKI.

13 Source: Brookings, "China's Influence on the Global Middle Class", October 2020.

14 Source: Refinitiv, "Bridging the global infrastructure investment gap requires better data", Sherry Madera, January 2020.

15 Source: TaiwanNews, "China's African debt-trap: Beijing prepares to seize Kenya's port of Mombasa", Duncan DeAeth, 27 December 2018: https://www.taiwannews.com.tw/en/news/3605624.

16 An English version of the report can be found here: https://flia.org/wp-content/uploads/2017/07/A-New-Generation-of-Artificial-Intelligence-Development-Plan-1.pdf.

17 Source: *The Economist*, "The Algorithm Kingdom: China may match or beat America in AI", 15 July 2017.

18 Source: Ant Group 2020 Annual Report, accessed March 2021.

19 Source: Pymnts.com "A Tale of Two Fraud Stats", August 2014.

20 Source: World Intellectual Property Organization, a UN initiative.

21 Source: NikkeiAsia, "China passes US as world's top researcher, showing its R&D might", 8 August 2020 - https://asia.nikkei.com/Business/Science/China-passes-US-as-world-s-top-researcher-showing-its-R-D-might.

22 Source: National Science Foundation, "State of U.S. Science and Engineering 2020", January 2021.

23 Source: ECNU Review of Education, "Accelerated Move for AI in China", Xiaozhe Yang, September 2019.

24 See Kai Fu Lee's AI Superpowers for more of a nuanced discussion on this issue.

CHAPTER 9

THE ECONOMICS OF THE FUTURE

"Education is our passport to the future, for tomorrow belongs to the people who prepare for it today."
—Malcolm X

Earlier we discussed why the present economic system won't continue to work without some significant changes. In this chapter we want to assess what specific changes need to be made for a future economy to function as it should. Particularly in respect to creating shared and sustainable prosperity.

This future is not as opaque as one might imagine. When we consider the possible trends and forces that will shape future economies, we get a partial picture; but when we factor in what must be done to achieve peak economic performance for humanity, it becomes much clearer. We've attempted to create prototype models to encourage the sort of thinking that is needed regarding the forces that we discussed earlier in Chapter 1. We start with a core assumption—an economy's purpose is to create happy and healthy citizens in a sustainable way—not just economic growth absent of a greater purpose. We conclude by explaining why this cannot be a zero-sum game, and what actions we think are required to eliminate long-term economic uncertainty.

Some of the ideas we explore have been raised by others in the past, and we hope to build on them in light of the big problems that need to be solved. We are also introducing new ideas with the aim of giving them oxygen to be discussed and studied. We are attempting to raise real solutions to the stickiest problems because, as Nobel Peace Prize laureate Betty Williams said: "There is no use talking about the problem unless you talk about the solution."

Without a global economic future that is inclusive and that provides opportunities for everyone to contribute and have a sense of purpose, we will remain divided as a species, ultimately leading to conflict and social breakdown. If that happens, it is certainly possible that a better system might emerge. But revolution and failed states are not elements of an orderly transition to an optimal state of human governance and productivity.

Let's take capitalism. Capitalism has been the dominant economic model for many nations for many years, but it is not the only model. It may not be the best model, or perhaps it is but it needs updating. To imagine that in a thousand years the present form of capitalism will be the very best system that can be developed to organize and reward human effort is a stretch. Changes to the economic system that we presently have are inevitable, it is just a question of how and when these changes will take place. So why not attempt to design something that works better now, why wait until the systems collapse? We need to shake off the shackles of previous economic thinking and policy and not be limited by the past.

The key to preventing a chaotic, dismal future is to plan for one that has the necessary pillars for success:

- equal opportunity
- inclusiveness
- a sharing of the wealth and abundance that the world can provide
- a focus on valuing the things that will advance humanity rather than the things that hold short-term transient appeal; and
- a system that respects the planet and resources we have as finite

No magic pill, no quick fix, but a path to optimal outcomes for humanity and the planet if the right path is taken.

Pressing economic issues

It is abundantly clear that the Global Financial Crisis (GFC) and COVID-19 accentuated failures in the present system. Today this still includes the quantitative easing that is spurring inflation, and the stimulus measures and the global "lockdowns" on travel and trade that will leave our children with a huge debt overhang that they will be dealing with their entire lives. Without the external shocks in 2008 and 2020, inflation and debt may not have been as problematic as they now are, but inequality and other issues would have remained failure points.

The pandemic has been an accelerant of some trends and issues, including the ascent of non-fiat currencies, the transition to digital-first industries, rising political and trade tensions, changes to the nature of globalization and how businesses and governments are planning for it, and leaps forward in AI, technology and connectivity. The dual crises have shown a need for regulatory harmonisation, to have more people remain in the workforce for longer, and—especially as AI takes hold—to have more technologically-aligned skills with labour working more with their heads rather than their hands. We need economies where the workforce is innovative and creative and not looking to the past for help with 21st century problems. Let's not forget that the deployment of AI has the potential to massively improve productivity and wealth generation too, but not for the benefit of all necessarily.

Growing wealth inequality and environmental degradation have been discussed as problems for decades now, but there is a sharpened focus on them in recent times—possibly because COVID-19 has led many people to think more deeply about such issues, and about what is of core importance—and, in line with the will of a majority of people, a stronger commitment to deal with both issues at a political level. The lack of political will and market ability to address these issues in the past is an indictment on humanity in some ways, but also evidence that the world's economies have never had to deal with problems of this magnitude.

As we've previously observed, for the digital "weightless" economy, the instruments in the old toolbox can't fix things on their own because they were designed for a different time with different challenges and capabilities. Time has run out on that playbook.

What happens if we don't change the system, and just double down on 20th century economic policy? Here are some of the broader challenges:

1. An era of damaging and hard to control inflation where monetary policy is no longer as effective as it had been
2. Declines in productivity
3. A disconnect between the largest economies as globalization stalls
4. Global debt levels that are completely unmanageable—if debt had been ramped up in response to, say, climate change it would have been one thing, but the levels of debt we have today in part have been incurred because of inaction and mismanagement[1]
5. The death of money as we know it
6. A rapidly worsening environment that results in a poorer quality of life and millions of avoidable casualties
7. An acceleration of income inequality due to technology shifts
8. A global population that is growing in size and living longer
9. A workforce whose process-oriented and problem-solving activities are progressively replaced by robots, algorithms and AI, leading to large-scale displacement
10. Acute labour shortages for the new jobs that emerge because of insufficient STEM education and immigration; and
11. With large scale techno-unemployment and the emergence of UBI, the potential loss of motivation and passion for those no longer employed.[2]

Let's dive into the apparent solutions, along with some radical ideas that could bring us closer to optimizing human potential and delivering sustainable prosperity as a global outcome.

Forces shaping the future economy

Aside from the blockchain and the digital economy that it has enabled, all of the trends and forces that we discuss in this section were evident in some form before the GFC. When Satoshi Nakamoto (or his creators) wrote that first whitepaper on bitcoin, little did we know that within a decade this would have created more than $1 trillion of economic activity. But then again, the 21st century would have likely required a rethink of money, markets, and assets in any case.

Blockchain: Data infrastructure for the 21st century

If data is the new oil, then edge computing, AI, and blockchain are the refineries and pipes of that oil. The invention of blockchain will be viewed historically as a major event. Blockchain made possible the rapid rise of digital money, including cryptocurrencies, central bank digital currencies, smart contracts, DeFi and NFTs. However, blockchain has much broader implications for pretty much every industry. Everything from energy to property to healthcare to government services to logistics to banking, will likely be disrupted and/or transformed by blockchain technology. Digitization of supply chains and commerce will need blockchain constructs to work securely without cybersecurity disruption. Consider just one aspect of the financial sector, fiat money, and what is predictable.

As previously discussed, fiat money gives governments an ability to manage how much money is in circulation in the economy[3]. This mechanism has created flexibility and relative economic stability for many years. It was easy to bank, it supported the growth of a multi trillion-dollar financial system, and it was easy for people to understand, borrow and use. As a trusted value exchange mechanism it worked, for its time.

More recently the prospect of an ongoing decline in the value of the dollar, and technologies challenging the adaptability, auditability and security of fiat currency, is pushing us to rethink how money itself should work moving forward. Printing more money is clearly not the answer.

> *"He had his cash money, but you couldn't pay for food with that. It wasn't actually illegal to have the stuff, it was just that nobody ever did anything legitimate with it."*
> **—William Gibson, Count Zero**

When asked to predict when the death rattle of fiat money will be heard, it won't be a specific year, but a waning of influence. Not because the death of fiat is up for debate—the future of money is clearly not fiat—but because there are so many variables in the mix and so many unknowns that it is impossible to predict precisely when the bell tolls. The future that William Gibson envisioned is one that we imagine as the most likely scenario—not that fiat is outlawed, just that better alternatives have taken over daily use.

We don't, however, underestimate the resolve of those who are wedded intellectually to fiat, nor the massive resources at their disposal. Likely there will be a longish downward trajectory for fiat as existing substitutes popularize, as technology evolves, as generational change takes place, and as time is bought by governments, regulators and policymakers, to allow for a transition to whatever non-fiat world they find most supportive of their political and economic ideals. But make no mistake, the core mechanisms of value exchange are already evolving.

Additional to its other advantages, blockchain technology is an innovation that will pump-up productivity, will revolutionize identity, healthcare and supply chain management, to mention a few.

Productivity as a future lever

Productivity improvement has demonstrated the ability to close income and wealth gaps, and in the past has seen distribution of the resulting gains more evenly across economic classes. For example, the Model-T Ford production line is credited with creating the proofs that led to the fastest middle-class growth we'd ever seen historically. Over time there needs to be convergence between advanced economies and emerging economies on productivity measures. Technosocialism requires it. Otherwise, the more advanced economies continue to accumulate wealth from productivity growth and developing economies will fall further behind.

Productivity's role in the 21st century economy

The commonly held view is that productivity is measured as output per input. Physical productivity is the quantity of output produced per unit of input, say, one hour of worker time. Multifactor productivity measures economic performance by comparing the value of output to the amount of combined capital and labour inputs used to produce that output. So, if a worker uses a machine to produce five items in an hour and each item is priced at $20, then the multifactor productivity is $100.

The move to digital-first industries and replacement of large portions of human labour by aglorithms is changing our understanding of supply and demand, and how productivity should be measured. The 21st century economy will mostly comprise intangible goods and services—at least in terms of value and spend. If demand increases, then supply could be simply processing cycles in a Gigafactory—managed by humans, but no longer dependent on human labour for productivity gains. How do we measure productivity when the input is, say, medical treatment that extends the working life of a person? Is it the value of a longer career? How do we separate the medical treatment that enables a longer working life from other changing variables such as work, relationships, our environment, or not abusing alcohol?

This is difficult to do, and questions like these have led researchers to consider what role social factors and cultural context might play in improving productivity[4]. Productivity itself may need to be redefined. Certainly, it will be interesting to observe what the productivity differences might be between places where there is collectivism and a sense of global identity, versus places where individualism and nationalism prevail, for example. If society is highly automated and we're living longer, the role of work itself is certainly going to be challenged. As will be our thinking on wealth and opportunity.

Both hard and soft infrastructure is needed for the 21st century. The workforce needs to have the necessary tools to work effectively and efficiently, but it also must be educated and healthy to be productive. And, as people become more productive, they should derive greater job satisfaction, investing greater mental energy in new ideas and innovation.

Planning for structural employment changes

In Chapter 1 we identified that one of the single biggest challenges for future economies is in the form of the technological unemployment that will occur as workers are displaced by AI, machines and technological revolution.

Automation in certain industries is going to create a boom, but it will certainly force the collapse of others; akin to how farm automation killed the agricultural sector during the early Industrial Age. What the overall impact of these changes will be on GDP is as yet unknown.

The second dramatic reframing of economies for the 21st century will be our response globally and nationally to climate change and its effects reshaping our planet. This will change economic priorities and create new industries and endeavours where huge opportunities for innovation exist.

These two factors will result in greater disruption in workforce management and employment than we've faced since the industrial revolution. One functional area of the economy where we are certain to see longer-term investment is in talent management. We will have more data and better processing capability to help understand and direct people toward doing the things that they do best. But we'll also be constantly adapting and retraining our workforce as jobs and industries evolve. Perhaps the most critical skill of successful people in the 21st century will be their adaptability.

This is a good thing because in the future economy people are going to have to work for longer. As life expectancy increases, medical treatments improve, and lifestyle factors change—resulting in people living longer and being fit and active for longer—we just can't afford to have people quit work and not contribute to society in their 60s or earlier. Society will expect that everyone is going to have to be productive for longer and contribute to society beyond what is today considered normal retirement. Millennials will be the first generation where having multiple careers during their lifetimes will be considered normal, and where ongoing learning, training and development becomes more critical. It may mean that there are age incentives to reward people who have done more and contributed for longer.

Let's take one possible scenario. If by 2050 typical life expectancy has reached 130, how long would you stay at school? Will it become normal for citizens to enter some form of national climate service for the first years of their careers to qualify for citizenship or universal basic income in the future? If we live longer we will have to start thinking very differently about the transition from student to workforce, from a single career to multiple careers, etc.

Regardless, successful future economies today need to prepare for an uncertain future, while attempting to tackle economic uncertainty for social stability. This requires investments in education, training, workforce participation (getting more people working), encouraging more women into the workforce, extending the age to retirement, as well as investments in infrastructure, research and development, opening up markets and increasing trade. Innovation and efficiency will be the core economic benchmarks that drive competitive performance. The goal of a structural increase in economic productivity, while creating fully sustainable economies, must underpin the operating philosophy of the 21st century.

Undoubtedly some of you are asking right now, who is going to pay for it? While the financial firepower of economies like the US has been leveraged to see those economies through global crises, only a small portion of that spend has been directly targeted at broad economic innovations. Thus, we are left once again with the prospect of using tax dollars to fund initiatives to renew infrastructure, retrain impacted workers, and fund programs that stimulate new R&D or technological advancements. The key problem with this strategy is we commonly see such tax revenue redirected due to political opportunism, with long-term planning responses suffering.

We need much grander visions of what is possible, and how economies satisfy that potential. GDP growth is such a limited view of the world. The most important element in the future economy should be humanity's viability and thrivability (as Aristotle reminds us). To that end, we need a new vision of what people will do in the 21st century. The key here is to understand the economic model that is most likely to succeed, and what individuals and markets need to do to fully participate in that future.

The Knowledge-Innovation-Creative economy

This book is about humanity and its future, so let's begin by discussing what we define as the Knowledge-Innovation-Creative (KIC) economy.

In the future, everyone will be a knowledge worker. Perhaps professional athletes and adventure tourism professionals aside, almost no-one will be paid for their labour in the later half of the 21st century. Creativity, initiative, critical thinking, humour, reasoning, judgment, insight, artistry, leadership, ideas, coordination skills, collaboration skills, management skills, research and investigation skills, will be in demand and all jobs will require some creative or intellectual ability.

This doesn't mean that everyone needs to understand quantum physics or how to calculate orbital rendezvous trajectories, and not everyone will need to be a master chef, but everyone who wants to work will be using brains instead of brawn to earn their digital currency. We are closer than many may think to living in a purely KIC economy, and humanity will be both liberated and challenged by it.

KIC work will be a great equalizer. Physical traits no longer matter, gender, ethnicity and age doesn't count; ideas, intelligence, and skills do. This should further empower women and minority groups, something that is long overdue and must be a part of strengthening the equity of future economies. It will also likely mean that jobs could be filled using curated algorithms that are blind to gender, race, height, weight, or any of the other factors that shouldn't matter now. Policy has as yet failed to produce these outcomes.

Removing biases in the shift from machine learning that simply replicates current process, behaviour and policy is an obvious prerequisite. Yet, that will be difficult because intelligence, critical thinking, and social skills will still matter. These human-differentiated skills will be rewarded, and not possessing those things will mean that a worker finds it more challenging to find work. However, the elimination of mindless jobs is a conceptual shift in the way we think about work. Some people value not having to think about their work and they engage in constructive apathy[5]; but not thinking won't pay in the future economy, so mindsets will need to change, or we'll need to support those who can't work in some way. We just don't get to keep mindless, human process-oriented jobs in the

21st century. For years we've trained humans to work like robots, but now robots will be doing the work of robots.

Technology dominance requires investment in people

Capitalism was designed to allocate dollars to the highest-returning investments over time. In the 21st century this means putting money into technology, not humans. Investing in AI will make sense for investors, corporations and markets, but leaves all other stakeholders wanting.

Creating economic incentives to invest in people and to value people is the way to balance our future technological advancements. Government incentives aside, this will also mean teaching people how to be ambitious about their own potential. But the biggest challenge is that education will need to be free and ubiquitous in this future—without this we will simply drive greater inequality and disparity. 21st century brands and companies will need to foster a culture where retraining staff is a core construct, rather than just replacing humans with automation. The good news is that much of this can be self-taught, if the motivation and incentive is there for that to happen.

Stop living in economies of the past

Knowledge, innovation and creativity have been important in every stage of human economic development. The difference today is that companies focused on KIC have demonstrated an economic value that far outstrips anything previously seen, and because of technology their products and services are ubiquitous, replicable, portable and oftentimes dramatically cheaper to reproduce. That's why we're constantly talking FAANG, BATX and other acronyms today when we talk about market returns.

The old supply and demand curve of yesteryear doesn't capture or explain the value dynamics of this world. The total cost of production doesn't necessarily correlate with demand changes—at least not for intangible goods. Economics simply hasn't kept pace with real-world dynamics. In the US, GDP calculations didn't even include software sales until 20 years ago (1999)! The metrics used today to measure wealth and prosperity are still way behind the mindset of many citizens on those issues. The previous chapter showed that we need to add factors such as happiness, well-being, health, longevity, satisfaction and similar into the mix when measuring the prosperity of nations and the effectiveness of economies.

Our take is that for most people:

future prosperity = f (tangible assets, intangible assets)

Where the portion of tangible assets will be greater for some and less for others, but for nearly everyone the portion of intangibles will exceed the tangibles over time. It is the nature of the 21st century.

As these core economics change, we won't any longer feel prosperous just by accumulating stuff. As economies compete with each other in meeting the basic needs of citizens through innovation, humans will naturally shift to accumulating meaningful experiences, with the realistic hope that they can and will accumulate more of those experiences while being healthy, mindful and pain-free.

No nation tomorrow can hope to become truly prosperous unless it develops this model of a KIC economy. The old guards of the Durant "pyramid economies" are going to find this a fundamental challenge to their existence.

Where true value lies

In the 21st century KIC economy, value is generated through the intellectual property that is embedded in high-value goods and services. The resulting products and services tend to have high value per weight or volume, and simultaneously—like AI and digital media—potentially no weight or volume at all beyond their external impact. The rise of this "weightless economy" makes the tyranny of distance as an obstacle to work location, connectivity, and the ability to generate and access value, a thing of the past.

The pandemic demonstrated that people will be able to live and work pretty much wherever they wish as long as they have access to some fairly basic tech. It also demonstrated that digital services like tele-medicine and remote schooling are possible, absent the structure of urban living. While most of humanity will still choose to spend much of their time in urban centres they will be able to have fluid lives and can be where they want as the mood takes them, without affecting their personal productivity.

Knowledge and innovation that generate wealth today are typically those that result in products and services that create value and are difficult to imitate or substitute. Knowledge that becomes easily widespread and cannot be protected or upgraded loses its value over time. Thus, future economies live or die on the ability to create and use new knowledge and innovations on an ongoing basis—ongoing innovation, not just static IP. This is where R&D is critical, thinkers and dreamers are highly valued, and risk-taking and forward planning are both essential traits.[6]

Being a KIC economy docs not just mean having "technology-first" industries and deep R&D, it also involves innovation in governance, coordination, cooperation, communication, business systems, business processes, standards, training and market development. It also means that IP or innovations we create that might benefit the world may be seen as greater inputs into the global economy than just simply the returns they generate to the creators.

There is no single metric that defines the potential of a KIC economy. Numbers of patents, the presence of leading companies, measures of educational achievement, and the portion of "knowledge workers" in an economy are all at best partial measures. How then will we distinguish the wealthiest nations of the 21st century?

Global Knowledge Index

The United Nations Development Programme (UNDP) Global Knowledge Index (GKI) 2020 reports on the "knowledge status" of 138 countries. The report on the GKI notes that investing in knowledge will further human development and wellbeing and that "addressing knowledge gaps in areas like the digital economy will be crucial to accelerate our collective efforts to transition to a zero-carbon future that protects and restores nature".[7]

Table 1 shows the top 10 ranked countries in the GKI 2020 along with a number of OECD nations that didn't make the top 10, including Australia, Canada, France, Germany, Italy, Japan, Korea, Mexico, Spain, Turkey and the United Kingdom; plus the two most populous nations in the world, China and India; four remaining countries that are among the 20 largest by GDP and included for that reason—Brazil, Indonesia, Russia, and Saudi Arabia; and then, for comparison, the bottom-ranked country Chad. For each selected country, its rank in the overall GKI 2020 index is given along with its knowledge index score out of 100, as well as its 2020 population and per capita GDP for 2020.

GKI Rank 2020	Country	Knowledge Index Score (overall)	Population 2020 (approximate)	GDP per capita 2019 (USD)
1	Switzerland	73.6	9 million	85,135
2	United States	71.1	331 million	65,134
3	Finland	70.8	6 million	48,678
4	Sweden	70.6	10 million	52,896
5	Netherlands	69.7	17 million	53,053
6	Luxembourg	69.5	0.6 million	115,481
7	Singapore	69.2	6 million	64,103
8	Denmark	68.3	6 million	60,657
9	United Kingdom	68.1	66 million	41,855
10	Hong Kong China (SAR)	66.8	7 million	49,180
11	Germany	66.2	83 million	46,232
12	Japan	66.2	125 million	40,063
19	South Korea	64.4	52 million	32,143
20	France	64	67 million	40,319
23	Australia	62.2	26 million	54,763
24	Canada	61.1	38 million	46,550
30	Spain	57.9	47 million	29,816
31	China	57.4	1.41 billion	10,004
32	Italy	56.6	59 million	33,090
42	Saudi Arabia	50.9	34 million	23,140
45	Russian Federation	50.6	146 million	11,606
57	Mexico	47.5	126 million	9,849
68	Brazil	45.4	213 million	8,755
69	Turkey	45.2	83 million	9,127
75	India	44.4	1.37 billion	2,116
81	Indonesia	43.3	271 million	4,136
138	Chad	21.5	17 million	707

Table 1: Global Knowledge Index (GKI) 2020: Scores, Population, and GDP per capita for top 10 ranked and various other countries[8].

The GKI rankings for 2020 show that the countries that do well tend to have relatively high GDP per capita. It would seem reasonable that the most knowledge-intensive economies are among the wealthiest economies. But wealth isn't the only factor. Proximity to other KIC economies, connectedness, industry focus, size and population also are noticeable factors, as is investment in education that produces knowledge workers, innovators and creatives.

Switzerland, with a relatively small population and ranking 1st in the GKI, borders much larger countries both in terms of population and overall GDP, such as Germany (11th), France (20th) and Italy (32nd). It has great physical connectivity. It also ranks highly for internet accessibility, ranking 9th globally in the 2021 Internet Accessibility Index.[9] A majority of Swiss workers are employed in the service sector, many of them in banking and insurance, which generates around 10% of national GDP, and being industries that deliver very high returns per employee. Switzerland enjoys the benefits of being able to service large surrounding economies, with positive economic spillover effects, from being focused on a few key high-value industries, from being super-connected, and from having a small population that is concentrated geographically.

The factors that position Switzerland so well in the KIC economy apply to many others in the top 10 GKI rankings for 2020. Finland (3rd) and Sweden (4th) share a border, have small populations, large hinterlands to service economically, and are in the top 12 ranked countries for internet accessibility according to the 2021 Internet Accessibility Index; they also have happier populations.

Singapore (7th) and Hong Kong (10th) have the advantage of having advanced economies in the fastest growing region in the world, and in the case of Hong Kong, China's position as the soon-to-be largest economy in the world has created massive demand and opportunity in its key industries. Hong Kong also ranks 4th in the 2021 Internet Accessibility Index[10].

The US and UK are different. For the US, its ranking in the GKI is in part due to being the largest and, in many ways, most advanced 20th century economy. Its massive resources have supported world-leading connectivity, and it has some of the most respected educational institutions and research centres globally. The UK has several of the world's leading universities, an education system at all levels that attracts large numbers of students from other countries and is highly competitive, and, as is the case for the US, it has the benefit of having been an advanced economy for a very long time—so the idea that education and knowledge is critical to being globally relevant has been culturally and socially embedded across several generations. These foundations give the US and the UK fundamentals to keep pace in the KIC economy, but only if sufficient investment is being prepared for the economy of tomorrow. Relying on 20th century industries or skills will be disastrous for both economies.

China (31st) does surprisingly well for a nation that only began opening up in the late 1970s, has the world's largest population, and a large land mass to cover. India, though it ranks in 75th place, also has the challenge of a large population that is spread over a wide geography.

In China, Shenzhen is the poster child of a smart city in the digital economy, Shanghai is a very international city with many of the world's largest MNCs having regional head offices there and at least 20 other cities have major industrial economies that have been steadily moving up the technology and value ladder bringing the people who live there progressively into the KIC economy. In India, the digital transformation is well underway in Bangalore and Hyderabad. India is the number one sourcing destination for IT-BPM services globally and has the second highest number of internet subscribers in the world, and the second largest tech start-up hub in the world.[11]

For China and India, the picture that emerges is one in which millions of skilled workers working in KIC industries are already available and concentrated in cities or regions that will advance quickly. As AI becomes more commonplace, this will enable large sectors of the economy to rapidly adapt, leading to changing economics that contribute to the prosperity of the nation as a whole.

For Chad, at the bottom of the GKI league table, and others in the same zone, their main hope in the short term is that wealthier nations invest in infrastructure that will help them overcome poverty and deliver their citizens a standard of living that every human deserves. The one benefit for these nations is they will have greenfield 21st century infrastructure that will be the engine of growth. At some point, as our technical capabilities progress, we need to recognize that we should support our fellow humans. The pandemic did achieve this result somewhat as we saw wealthy nations like the US supply vaccines to worse hit poorer nations. This is the right thing to do, and shouldn't be an economic consideration in an enlightened world.

But there is no quick fix, and herein lies the problem. Even if the needed investments were being made today, it still would take a couple of generations for citizens of the very poor nations to transform their own economy, and wealthy nations who measure success in political and economic terms might retreat because the ROI isn't there yet. For the future economy to work for all of humanity, and for it to overcome the biggest challenges we face, everyone needs to pull together and feel like they are part of the same competition, not that they have been relegated and marginalised to the extent that their contributions aren't valued and don't matter. We clearly need economies that create competition for the purpose of advancing all humanity, not that foster competition against each other. That's the core philosophical shift of technosocialism.

A checklist for the KIC economy

The number of KIC workers indicates that the KIC economy is still in its nascent stage. Rather than developing naturally, it's likely being constrained to some extent by traditionally-held economic practices. For an economy to become a world-class KIC player, these are the elements that must first be in place:

- **The Right Skills**—skilled STEM and creative professionals are essential for a KIC system, not only for the generation and application of new knowledge, but also to being able to use and adapt knowledge produced elsewhere. Right now economies like the US, UK and Australia are likely to have both unemployment due to the impact of AI, along with labour shortages due to lack of adequate education for KIC development.
- **Job-Ready Professionals**—some softer skills are increasingly important in the KIC economy, such as teamwork, analytical

problem solving, communication skills, entrepreneurship and leadership. This requires more exposure to real business scenarios through internships, etc blended into educational curriculum. Germany's apprenticeship programs that combine education, job training and work experience are better preparation for KIC jobs than typical western university curricula.

- **Continuous Learning**—students in university today are expected to have somewhere between 3–10 different jobs in the first 10 years after graduation. A significant portion of our time will need to be spent on ongoing education and development, and this will need to be supported by companies and government, particularly as automation accelerates.

- **Broader Participation**—increases in the mature age labour force participation rate and in female workforce participation, by even modest percentages, would bring about transformative increases in GDP in most nations. This could be achieved by increasing the pension age and restricting access to pension funds until workers reach that age. Making childcare more affordable and giving women greater tax incentives would encourage their long-term participation also.

- **Brain Drain versus Immigration**—countries cannot afford to lose their best talent, and yet throughout the 20th century this has often been the case for countries that have been developing skills in global demand[12]. The emergence of "working from anywhere" culture may also provide a foundation for our best and brightest to remain at home, while working for global initiatives. As we discussed in Chapter 6, capturing KIC talent is going to become super competitive in the 21st century, with massive incentives being offered by many countries to attract smart talent not only to advance the economy but to grow consumption and participation as birth rates decline.

- **The Innovation Mantra**—innovation should be the lifeblood of the KIC economy. No single entity or individual can produce all the key innovations needed in the 21st century, so the next best thing is to build on other's success. Indeed, we think in respect to things like climate change mitigation that IP laws should be suspended, or inventors should commit to making their core inventions open source, as Tesla has done with a number of their patents.

- **Unicorn Universities?**—building commercial collaboration in innovation or prototyping labs could incentivize and reward more researchers to do work that can be rapidly commercialized. This could lead to faster ideation and clear employment paths, along with solving the funding problem that many universities face. This would stop the cycle of dumbing down the curriculum to compete for the limited pool of fee-paying students. We must commercialize the research that is done in universities, create incentives to attract students interested in research and development, build tax incentives for companies to fund these programs and draw in industry leaders and mentors who wish to energize and inspire the student population. Scaling this capability through digital platforms may also deliver us the first global unicorn university.

- **UBI and R&D**—KIC economies need to create engines for continuous R&D investment. This is why we think Universal Basic Income that allows individuals to explore their creativity and ideas without the basic concerns of funding would be a boon for wealth and GDP growth. There needs to be a co-evolution between business and humanity. AI will help us upgrade and advance humanity. But as AI makes life easier and more comfortable there is a risk that people will get bored and restless. The evolution of AI must happen in a way that allows us to generate more meaning and purposeful endevours for all.

- **Remove Capacity and Technology Constraints**—subsidies for 20th century infrastructure should be eliminated in favour of technologies and capabilities that will accelerate our global advancements as a species. For example, subsidies for fossil fuels should be immediately removed and that money invested in renewables and carbon neutral infrastructure. In this way we could transform energy grids at 10 times the speed of the free market.

- **National Incubation Hubs**—Silicon Valley, Shenzhen and a few other places aside, there is not enough investment to support sufficient national innovation, creativity, technology development and new entrepreneurs. One solution is to create national incubators for KIC talent that are free of red-tape, and that are funded with oversight from those with experience in the sectors rather than bureaucrats. National incubators could be linked to similar incubators in other countries to encourage competitive cooperation and cross-fertilization of ideas in ways that benefit all involved, including the sponsor nations.

Globalization, nationalism and collectivism

Ongoing globalization has enormous implications for the dynamics of the future global economy. But it is becoming more fragmented and murkier than it was a decade ago, and less assuredly will it lead to the sort of positive global connectedness that was envisioned at the turn of the millennia in 2000, when trade grew to nearly half of global GDP and hundreds of millions of people achieved middle class status by participating in the global economy.[13] The pushback against globalization has hurt most economies.

The risk that the globalization trajectory that gathered pace in the 1980s and continued with gusto pre-GFC will limp and falter is growing as nationalism rises, as trade wars become more entrenched, and as broad multilateral trade agreements are shunted aside in favour of bilateral agreements of unequal benefit and by insider-club multilateral arrangements.

Some view China's Belt and Road Initiative as an attack on other's national and economic soverignty[14], others view the former Trans-Pacific Partnership (which excluded China) in the same way. Organizations like the Five Eyes intelligence alliance has been criticized for provoking an escalation in conflict rhetoric, for impeding trade, and for being anti-globalization also. The pandemic dealt a further blow by promoting reductionist nationalism and, correspondingly, the belief that reliance on others is bad, self-determination is paramount, and, therefore, doing as much as possible domestically is safer.

Nationalist policies assume economic principles stay relatively static and commerce and collaboration based on the internet will cease to be effective. That's clearly fake news, so instead we must ask the question: who is going to get flattened and who is going to do the flattening as ideas, talent, and activities continue to be sliced and diced and scattered around the world?

The economies that dominate (flatten opponents) require knowledge of technology and global markets, an ability to rapidly attract top echelon knowledge workers, and an ability to spot emerging trends or create trends, while at the same time having access to global sources of production and support and being able to coordinate everything. They become the super apps of the world's economies—central hubs of knowledge, innovation and creativity bringing in talent, investment and startups from around the globe, and services that drive global consumption and participation.

Obvious candidates to be flatteners include the US, China and most developed nations, at least to some degree. However, nations like Indonesia, for instance, a nation identified as one of the "top three digital rising economies" and supported by initiatives with long-term vision such as the "1,000 start-ups movement" and investments in innovation and entrepreneurship[15] could hope to carve out a niche that keeps them relevant.

AI should be seen as core infrastructure, as should the ability to start a company remotely for the purpose of attracting talent and investment. Innovation and incubation hubs that work for talent on the ground as well as remotely also are important. Think of these "born global" companies choosing a global base much in the way a company might have chosen an office building in the 20th century. You'll locate your remote global business in whatever state provides the government support, packaged business services (smart banking/legal/accounting/lending) and pipeline of talent and global clients, all with the smart ability to dispatch, bill, pay and deploy.

Of course, there is a risk that globalization will create winner-takes-all markets in which the nations that create the ideas and own the intellectual property capture nearly all of the value that is created. Those countries that are flattened will be relegated to being idea and knowledge takers rather than makers and will struggle to develop the systems that make them competitive in terms of talent or infrastructure. There may be other advantages that these economies can leverage, however, such as natural resources, coveted tourism and living locations, or abundant produce that can be exported.

Smart Cities as global hubs

Cities are the nodes that connect national markets to international markets. The further a city is from other major global markets, the more important it is as a connector to the global economy. Achieving innovation means innovating in cities with the largest populations, where they are globally connected, where knowledge is concentrated, and where universal services provide a superior quality of life.

Cities that are far from large markets tend to perform their functions less well than cities in larger countries with denser populations, or that are next to larger countries with whom they trade and interact on other dimensions. So, Sydney, for example, does not have the scale or scope of cities like London, New York or Hong Kong. But while it is not possible to change the scale of Sydney or its economy substantially in the near-term, it is possible to overcome some of the local barriers such as city structure, urban planning, lack of infrastructure investment, lack of world-class connectivity and world-class ICT, and a lack of high-level skills, that prevent a city like Sydney from reaching its full potential.

Smart cities are far more relevant, competitive, and economically important for future economics. Their citizens enjoy a higher standard of living. For any city to be a world-class smart city it must be powered by sustainable, renewable, smart, digital innovations and inclusive mechanisms that work for its citizens.

Smart cities will be places that people want to live and work in. They will have financial, social, cultural and environmental capital. They will attract the leading global talent and this will reinforce their position as being attractive places to live and work.

Ranking today's world cities

The Globalization and World Cities Research Network (GaWC), created in the geography department at Loughborough University, UK, describes "world cities" as being those that are best linked to the rest of the world as indicated by the locations of headquarters of multinational companies, the presence of offices of major multinational professional service firms, and how linked they are with the world economy. The top 20 connected cities according to GaWC in 2020 are in the table below.

Rank	Country	Rank	Country
1	London	11	Los Angeles
2	New York	12	Toronto
3	Hong Kong	13	Mumbai
4	Singapore	14	Amsterdam
5	Shanghai	15	Milan
6	Beijing	16	Frankfurt
7	Dubai	17	Mexico City
8	Paris	18	Sao Paulo
9	Tokyo	19	Chicago
10	Sydney	20	Kuala Lumpur

Table 2: Top 20 connected cities according to GaWC in 2020. (Source: World According to GaWC 2020, Globalization and World Cities Research Network).

Table 2 shows that China has three of the top 10 connected cities. No other country has more than one. Four out of the top six connected cities are in Asia, further demonstration of the continued rise of Asia and its importance.

Only one of the top 10 connected cities is in the US, though LA is in 11th spot and Chicago at 19th, and Australia (Sydney, 10th) and Canada (Toronto, 12th) each only have one city in the top 20. Those are very big countries to be connected in a globally competitive way through one city, suggesting that those cities must remain focused on being at the forefront in developing infrastructure and attracting talent.

Right now, the smart money would be on the 20 cities in Table 2 being leaders and winners in the future economy—in some cases, they will win even if large portions of their home country do not advance to the same extent. The challenge for countries in that position is to draw on the strength of their connected city to develop their overall economies in a way that reduces wealth and income inequality. Failing this, they will find that tensions between the city and other areas spiral, leading to social unrest. But Smart Cities will reshape the competitiveness of nations throughout the 21st century[16].

Interconnected economic regions

As the challenges of humanity become more interconnected, so do economics and policy. While some local issues of infrastructure, resource management and services are manageable within current borders, others, like pollution, water quality, and climate action, require greater collaboration and joint policy development and investment.

For Hong Kong, which suffers the effects of pollution from cities north of the border, activity that improves air quality needs to be regional. For Bangkok, Ho Chi Minh City and Phnom Penh, the cloud of smoke-heavy smog that comes from cane sugar farmers in Laos and northern Vietnam aren't something they can fix on their own. The deforestation of Indonesia and Malaysia for palm oil, and the smoke from fires that pushes into surrounding nations, is an international policy issue. The flow of capital must be constrained by better policy for our neighbours—this means surporting technologies and industries that incentivize polluters to work within the region to improve the quality of air and life.

Through communication and knowledge sharing across cities in a region, activities such as managing traffic, parking, storm warnings, accident reporting, street lighting, public facilities management, e-government services, utilities planning and management, GPS tracking of public transit and related activities could be collectively managed. With sufficient foresight, it might be possible for neighbouring cities to work together to develop compatible programs that allow for maximum sharing of knowledge and experience, and cheaper systems and infrastructure development.

Environmental, Social and Governance (ESG) objectives reshaping markets

Disillusionment with the present economic system, with elements of pure greed-driven capitalism, and dissatisfaction with social conflict and entropic selfishness by a privileged majority have created multiple national and corporate movements that tie investments of time and money—particularly taxpayer money—to Environmental, Social and Governance (ESG) objectives.

Sustainable finance is the practice of integrating ESG objectives into investment decisions and operational frameworks to bring about sustainable development outcomes. This includes mitigating the adverse effects of climate change, and much of the heightened interest in ESG is due to awareness of the effects of climate change. Banks like Crédit Mutuel, UBS, Sberbank, State Bank of India and Intesa Sanpaolo have all reduced their investments in fossil fuel financing by more than 50% over the last five years in parallel with the Paris Accords, as their boards have committed to core ESG objectives like carbon neutrality.

> *"In 2018, we made a decision to stop all financing for coal-fired power plants and coal mining in all countries."*
> **—Crédit Mutuel, a French cooperative bank**

Standard Chartered in 2018 made the same public commitment to cease all fossil fuel related projects, but was exposed in 2021 by the *Financial Times* for providing part of a $400 million syndicated loan to Rio Tinto, one of the world's largest coal miners[17].

A series of landmark actions in the boardrooms of Big Oil and across courtrooms globally have already started to have a strong impact on the world's largest energy resource companies. ExxonMobil, Chevron, and Royal Dutch Shell all faced unthinkable actions against their core operating principles early in 2021. Shareholders at ExxonMobil accepted shareholder action led by a small hedge fund in overhauling the board. Investors in the US defied Chevron management in respect to a pivotal vote around climate response. And a Dutch court ordered Shell to drive down it's global carbon emissions in line with the Paris accord. This all happened within 24 hours[18]. Actions by shareholders against corporations and their boards who drag their feet on social consciousness will continue and get more vocal and transformative over the next decade.

"Impact investing" is an investment approach that aims for a more sustainable future for society while also delivering acceptable financial returns. Again, climate change is a driver of impact investing, as is a desire to address income and wealth inequality. Most impact investing is done by institutional investors such as hedge funds, private foundations, and pension funds, but increasingly socially conscious investor networks and financial services firms have sprung up, enabling more individual investors to participate. This trend is growing as gains in the "democratized investing" space occupied by those in the Reddit investor army, for instance, hit the headlines and give a sense of egalitarian camaraderie to the foot soldiers investing along ESG lines.

The filter now for sustainable and impact investing is whether the project or investment is a net positive or negative for society, not simply the returns it will make. Over time, this filter has widened in response to concerns about a range of social, environmental and community-level factors. This has changed the way that public policy is framed and implemented in many parts of the world, with many public sector investments now required to take social, environmental, and economic impacts into account in deciding whether they are compatible with expectations regarding purpose as well as profit.

Into the future, particularly as trade and globalization expand and workforce mobility increases, catalytic benefits will be more of a focus, as will overall effects on wealth distribution and income equality for any given spend or investment.

The World Bank was early in indoctrinating this thinking into its work in allocating resources, stating in 2005 that World Bank-funded projects should be assessed for their ability to reduce poverty, and that "…ethnic, gender and racial inequality are dimensions of as well as causes of poverty [so] it is also necessary to assess the distributional effects of an investment, or a change in policy, on these groups"[19].

Broadening the idea, the notion that investments should do no harm and ideally do good beyond merely making money has become widely accepted in the corporate world because it is ethical, and because it also makes money.

From fringe to mainstream

For a time, investing along ESG lines was seen as a fringe endeavour. Something nice to have for press releases. In fact, up until recently you wouldn't see this as a core mission statement, just something the PR department could push out for feel-good coverage. But this mindset is changing along with the size of the impact investing market, which is now estimated to be $715 billion and growing rapidly. In fact, the International Finance Corporation (IFC), part of the World Bank Group, has estimated that the impact investing market could be as large as $26 trillion over the coming years if the right opportunities were available[20].

Citing a change in Millennial and Generation Z behaviour toward a more "conscious brand of capitalism", The Deloitte Global Millennial Survey 2020[21] found that 74% of respondents intend to take actions to have a positive impact on their communities; half of them believe that it is too late to repair the damage caused by climate change[22]; and a majority are prepared to punish companies whose values conflict with their own.

To quote one recent MBA student who participated in IFC research: "Our generation knows we have to change what we're doing, because our parents' capitalism has led us to where we are now, and it's just not sustainable."[23]

In 2020, a survey by Global Impact Investing Network found that 88% of respondents had achieved financial returns that were at or above their expectations, and 99% of them felt that their impact expectations had been met or exceeded[24]. So, the idea that investments can both do good for society and be good for the investor has been proven in practice.

And the shift in thinking that links social outcomes to employment and investment is happening at the highest levels. In April 2020, BlackRock, the world's biggest asset manager, launched its Global Impact Fund. On climate change, US President Joe Biden said in his first address to Congress:

> *"For too long, we have failed to use the most important word when it comes to meeting the climate crisis: Jobs. ... For me when I think about climate change, I think jobs ... The American Jobs Plan will put engineers and construction workers to work building more energy efficient buildings and homes...Electrical workers installing 500,000 charging stations along our highways."*
> **—President Biden on US infrastructure and Climate Response**[25]

GLOBAL GOVERNANCE

Presently no organization has global executive, military, or legal power over all citizens of planet Earth. But climate change might just create the need for much greater coalescing of governance and response.

The science-fiction author Kim Stanley Robinson recently wrote on the creation of "The Ministry of the Future"—a fictional UN, World Bank, OECD and WHO led department that has widespread powers to enact policy and operationalize actions to prevent the worst elements of climate response as hundreds of millions of people start to die across the planet.

A supranational union also existed in Aldous Huxley's imagined "World State", governing over a dystopian society that is defined by its intelligence-based social hierarchy[26]. This is but one of the better-known fictional attempts to portray a world order that is more ordered and stable[27] than the one we have. Most are great as fiction, but we'd be wise to avoid putting them into practise.

There have been attempts in the real world, too. Skipping to modern times, President Theodore Roosevelt spoke of a League of Peace to be formed by the "Great Powers" to avoid conflict among themselves and also to prevent the peace being broken by others[28].

The League of Nations, a precursor to the UN, was formed in 1920 with the mission to maintain world peace and having at its peak 58 member nations before folding in 1946 following the evident failure of its mission to prevent war.

The World Federalist Movement of the 1930s and 1940s, whose supporters included Martin Luther King J., Albert Einstein, Winston Churchill and Mahatma Gandhi, advocated for global democracy subject to principles of solidarity and subsidiarity, and gained traction among those who sought to prevent future war but had little faith in the UN being the right organ for that.

The idea of having some form of global government has been around for a long time.

At present, the closest we come to having an organ with global hegemony is the United Nations (UN), which has 193 member states out of a possible 195 in the world[29]. There are many multilateral organizations that have been formed for a specific purpose, such as the International Monetary Fund (IMF), the World Trade Organization (WTO) and the Organisation for Economic Co-operation and Development (OECD), but none compares to the UN in terms of membership and global participation.

The UN came into existence in the aftermath of World War II with the aims of preventing another world war[30], promoting peace, protecting human rights and upholding international law. In recent times the UN has authored a moral playbook for humanity in the form of the Sustainable Development Goals (SDG) for 2030, that all UN member states agreed to in 2015, to address global challenges including inequality, climate change, poverty and environmental degradation[31]. The SDG also calls for action to limit global warming.

The UN is vibrant proof of the ability for global cooperation on the issues that are of the greatest importance to humankind. For this reason, the UN is the logical foundation on which to build a new form of global leadership with a participatory and egalitarian governance. Our common cause is climate change and environmental threats, poverty, and inequality; any one of which is a threat to all humankind, and all of them being challenges that the UN is already working to tackle.

Article 25 of the UN's Charter states that "The Members of the United Nations agree to accept and carry out the decisions of the Security Council in accordance with the present Charter"[32]. This limits the power of the UN to the scope of the present charter. Perhaps amending the UN Charter to have wider and more expansive powers for global action and enforcement on issues pertaining to all humanity is the answer, or perhaps that falls short and more is needed. But it would be a start. It is also a reminder that for any meaningful action to be taken by the UN, the members of the Security Council must first agree.

Presently, there are 15 members of the UN Security Council (UNSC). The five permanent UNSC members are Russia, the US, France, China and the UK, and they have veto power.

Article 27 of the UN Charter states that all permanent members of the UNSC must vote in favour of a substantive resolution for it to pass. Therefore, by not casting an affirmative vote a state can exercise unilateral power to prevent the passing of a resolution. The veto creates consensus through compromise, thereby satisfying the liberal ideal of international cooperation. However, the veto's efficacy is diminished by an anarchic world order. Permanent UNSC members may use their veto power to safeguard external sovereignty. The veto gives them control over outcomes that can be used as a bargaining chip in their interactions with other states. The veto helps them further their own national interest and promote internal sovereignty. This goes against the liberal values on which the UN was founded. The UN should address transnational issues as a product of globalization. This requires pooling of state sovereignty and interdependence between constituent states for the global good.

Global power dynamics have changed since the end of World War II, yet the five permanent UNSC states were given the veto-power because of their victory in that war. Through this lens the veto is an anachronism. It is contested whether the veto creates enthusiastic consensus or forced agreement. The reductive nature of the veto can simplify decision making, but it can also create tensions between states holding veto rights, which may prove counterproductive to future decisions.

One issue with the UN, as was the case with the League of Nations and with most of the ideas forwarded in support of some form of global hegemony, is that they all were born of war, fear of war, or a desire to prevent war. Conflict was the epicentre of their birth and the crucible in which they formed. But the challenges of today are quite different to those that confronted humanity when the UN was formed. This suggests that if the UN is the model for designing a future global government, then its postulates need to be modernized.

Critics of the UN say that it is bloated, inefficient, bureaucratic, unwieldy, slow, undemocratic, unduly influenced by the larger members, and that it is good at setting goals but poor at achieving them[33]. These issues would need to be addressed and there would be challenges to overcome if the UN or an organ that evolves out of it were to be given even greater powers.

Comfort that change is possible is demonstrated in the form of efforts by members of the G4 (Brazil, India, Japan, and Germany) to prompt reforms to the Security Council by offering to initially forgo veto powers if admitted as permanent members. This is indicative of a willingness on the part of powerful nations to compromise in some areas.

Perhaps this attempt at reform is a sign that the limitations of the veto can be surmounted, and that other changes are possible that would lead to a new form of global cooperation. If that hope proves to be justified, we propose that the UN be given new powers and a broader remit under an updated charter. Veto powers for the permanent members of the UNSC should be removed and the past intransigence of the members must change, particularly in the light of present existential threats. The new powers would in the first instance extend to dealing with climate change and infectious disease. A logical next step would be for the new UN to have control over nuclear weapons with a focus on monitoring existing weapons and stopping nuclear proliferation, with the massive sums presently being spent on nuclear weaponry being redirected to fight climate change.

Some will forever believe that the idea of global government is quixotic and unattainable. This is why we suggest taking incremental steps that build on what the UN already does; proof of concept, next step, then prove that one, and so on. We need to step away from a Foucauldian world in which the relevant knowledge is concentrated in the hands of a few and all the financial firepower is controlled by them. If we don't then inequality will worsen, the environment will continue to erode, and the chasm between the empowered and the disempowered will grow—for all the good that it will do the empowered when they preside over an uninhabitable planet full of discontents.

We don't need a global government to fix the problems of the world that are emerging, but we need global governance that is cooperative rather than combative. This doesn't require surrendering national interests to institutions controlled by foreign nations. This means good species-wide cooperation on strategies that require commitment of massive budgets globally to respond to the changes thrust on humanity. The good news is, that as we automate much of this governance and regulatory function, it simplifies and radically reduces the cost of these mechanisms.

Perhaps the Ministry of the Future is not so far fetched.

The Asian century

The rise of Asia-Pacific economies has for decades been one of the most significant forces shaping the global economy. As Asia's share of global GDP increases to match its global share of population, resulting in nearly a doubling of its present GDP position, markets in Asia will become more important for all goods and services, likely both as producers and consumers, and Asian demand will spike massively. Three out of five of the world's largest economies will be in Asia by 2050. In some areas, Asia already outcompetes the rest of the world. As Asian immigration and investment continue to create even closer global ties, and as Asian nations become more affluent, they will compete across all arenas of the global economy. It is with good reason that many view the 21st century as being the "Asian century".

This brings with it a geopolitical rebalancing and demographic shift, with significant implications for changes in consumption. It means that without leadership and support from Asia-Pacific nations, no major global economic project or idea is likely to succeed. The good news is that if cooperation and agreement is achieved among countries, then these markets represent substantial new opportunities for companies from the rest of the world.

The opening of Asian economies, improvements in transportation and communication, international flows of knowledge and technology, modern logistics systems, and the evolution of major multinational companies that slice up their activities into finer and finer parts and place each part in its optimal location have spread economic activity around the world to the point where shares of global GDP could revert to shares of population, at least for countries that are sufficiently open and able to enter the global economic system. It is a safe bet that Asia's trade and economic development will only become more important over the coming years, so it would be wise to have a strategy for engaging with Asia.

Strategies for engaging with Asian economies

The view is often expressed that the likes of the G20 nations have an advantage over many countries in Asia because of their more advanced economies. If so, then presumably they should be able to sell high value-added goods and services into Asia as economic prosperity improves there and middle-class growth balloons. But for many economies, policy and cultural bias will dramatically inhibit their collaborative potential with Asia. Economies like the US, UK, and Australia will need these strategies to engage with Asia:

- *Need for an Asia-Pacific mindset*—believing that there are opportunities in Asia and seeking out and exploiting those opportunities are very different things. If businesses are to take advantage of opportunities in Asia, then they will need to be proactive in understanding and addressing customers, partners and competitors in those markets. Mindsets need to shift to understand that in many areas Asian economies are already outstripping the biggest economies of the 20th century. The stereotyping in American media of China as a place that lacks innovation and copies the US is an example of a mindset that is outdated and a barrier to economic cooperation.

- *Need to educate about Asia*—there needs to be a focus in education on Asia literacy. Relatively few Westerners study an Asian language, and knowledge of the political, economic and social systems in Asia are severely lacking as well. Having an "Asia-ready" workforce requires that the skills that are needed to do well in Asia are taught.

- *Asia needs to be better understood*—integrating more closely with Asia requires an understanding of the cultures, societies and business practices. Businesses need to understand that the markets in Asia differ greatly in almost every dimension to those outside of Asia.

- *Employ talent with Asia experience*—very few company directors and senior executives in countries outside of Asia have direct experience in Asia[34]. The problem is compounded by the fact that few recruiters outside of Asia have a detailed working knowledge of Asia and are not well placed to identify the talent that is needed to help companies outside of Asia do business there.

- **China and India are important, but the rest of Asia is, too**—much of the present discussion on the Asia-Pacific region focuses on China and India. While they are the largest and, in many ways, the most valuable players, there are a great many opportunities in other parts of the region, and these should not be ignored. The combined weight of the ASEAN economies, the enormous potential of Indonesia, the enduring economic strengths of Japan and South Korea, and the economic booms in Vietnam and Thailand are all hugely significant.

- **Understand and leverage free trade agreements (FTA)**—many existing free trade agreements between countries in Asia and the rest of world are poorly understood and, therefore, not fully capitalized on by companies. Striking an FTA should do more than positively affect thinking about the possibility for commercial relationships, it should enable real and immediate business opportunities, otherwise companies won't pay attention and the promise of an FTA will not be realized in practice. There is a wide range of expertise in Asia that can be leveraged to help western companies penetrate Asian markets.

- **Adjusting for differences**—Western democratic nations have very different social, political and cultural systems to those in place in much of Asia. They have different internets, too. These differences need to be understood and, at some levels, accepted if Western nations are to do more business in Asia. Common ground needs to be found in order to optimize trade, to develop the future global economy for all, with the required global and regional policy changes and responsiveness.

- **Encourage foreign investment**—there is huge scope for foreign investment and the opportunities that exist should not be derailed by the few instances in which there have been challenges and problems in the past. Most nations would find that investment from overseas is beneficial to their overall development, and it would be a shame for political or vested interests to derail programs for the wrong reasons. For example, recent US attacks on Huawei, while grounded in a national cyber security debate, are logically just as much about Huawei's massive technical dominance in the areas of 5G and Edge Computing.

Global regulatory reform

While regulation is necessary, it is important that regulations are sensible and that they do not needlessly shackle businesses. Much of the laws that govern the modern world are centuries old and built for a very different time. Reforming global regulation will require intergovernmental cooperation and consultation with business and the community, as well as insight into what comes next for the economy, the environment, business and the needs of people. Regulation should be an enabler of potential rather than a limiter of it. In many ways regulation today is designed to reduce system risk in the economy, but it does so by enforcing a very inflexible and historical view of how industries should operate. For example, in the United States the key act of law that governs financial inclusion is the Community Reinvestment Act (1977). Over the last decade we've learned that the CRA accentuates financial exclusion today and that the mobile phone has done more for access to financial services than any laws that nations like the US put in place 50 years ago.

Increasing globalization means that there is a need for global regulation, but more critically as we automate governance both nationally and globally—encoding regulations in AI will mean a studied and purposeful revamp of laws on the books. It will require dramatically different skills such as ethics, machine language bias filters, and core regulatory and compliance technology infrastructure. Other forces driving the need for worldwide regulation include the globalization of the professions and their governing bodies, and the activities of various multilateral organizations such as the World Bank, IMF, World Trade Organization, European Union and various United Nations organs.

Many of the drivers for changes to transnational regulation(s) were in place before the global recession in 2008 but, in the light of two of the biggest economic crises of the last 100 years, those same drivers are becoming greater catalysts for change. Beyond the GFC, other drivers of transnational regulation include cross-border technology and trade development, increased mobility across markets, organizations that use technology to scale globally such as social media platforms and super-apps, companies being listed on multiple exchanges, and changes in risk assessments by companies and governments.

Artificial Intelligence stands out as one obvious focus for future regulation that could be treated in this way. Global cooperation could avoid needless cost being embedded into the regulatory system and would, in any case, very likely lead to better and more effective regulation around AI. In areas like combating financial crime and money laundering, the only effective regulation is transnational, as should be the case for arenas like immigration, passport and border control, space exploration, climate mitigation and automated transportation and supply chain systems.

Starting now, we suggest that new regulations could be evaluated in terms of impact assessments to measure their effectiveness and that the OECD's Regulatory Impact Assessment (RIA) provides a potential model for doing this.[35] Lessons also can be learned from nations that are presently best at finding balance and in lessening the burden of government regulation. At time of writing, the most recent World Bank data on Burden of Government Regulation had Singapore as being the least burdensome, while Venezuela was most burdensome out of 136 countries for which data was available[36]. We note that Singapore also has an integrated plan for becoming a smart nation, and that it is way ahead of many other nations in becoming a digital economy, in developing a Digital Government Blueprint, and in creating a society for digital inclusion. Benchmarking Singapore and other leaders for regulation is one way for other nations to make progress.

As is the case for so many other issues, we are at an inflection point. We don't underestimate the challenge of achieving a consensus on regulation at a global level, but this must become a foundational commitment for species-wide protection. If we continue to drift along and bear the massive cost of the regulatory divergence and fragmentation that is symbolic of a divided world, we risk slipping into chaotic exclusionary division. If the largest economies join together to lead the way on regulation—with a focus on ensuring that it is progressive and effective—then others will follow. In fact, others will join the global regulatory infrastructure we create.

PROPOSAL: A GLOBAL CORPORATE TAX

President Joe Biden has proposed a new global minimum company tax rate for multinational companies of at least 15%, with a preference for a higher rate of 21%, in order to close the gap between the tax that companies in the US will pay and what multinationals outside of the US would pay if his proposal that the US corporate tax rate increase to 28% (up from 21%) is approved[37].

This is not the first time that a global company tax has been suggested—the OECD and others have been calling for it for some time[38]—but having it in the presidential prism raises the bar and means that it will be more seriously debated and considered than ever before. In financial parlance, the tax problem is identified as "Base erosion and profit shifting" (BEPS). It is a global issue and fixing it requires global cooperation. Perhaps it's time has come.

The heat on companies that are accused of shifting profits to low tax environments to reduce their tax liability has intensified as the digital economy has blossomed. This is because in the digital economy there is no need for a factory or other form of risky capital investment to be placed in a particular low tax rate location to earn revenue that is taxed there rather than in a high tax jurisdiction. Instead, the intellectual property can be placed in the low tax location and then revenue from a high tax location can be directed to the low tax location under licence or operating agreements that take the taxable dollars to the low tax environment.

The tax benefits from this type of transfer pricing might be good for the company and its shareholders, and it might be legal and clever, but shifting taxable dollars from high tax locations to low tax locations has the effect of reducing the tax base in countries where most of a company's activities are performed and where most of their staff live and work. The social good in terms of education, healthcare, infrastructure spend, environmental protection—all the things that are funded by the public purse—is undercut because the funds available for those items is smaller than it should be. The OECD estimates that BEPS is costing governments somewhere between \$100–240 billion in lost corporate income tax revenues per year.[39] This amount will increase as the KIC economy marches on and as the digitalization of business makes up an even greater portion of global GDP.

Clearly this is something that those countries that miss out on the tax dollars are likely to be unhappy about. And this activity is engaged in by some of the biggest and wealthiest companies in the world, companies that many feel are not paying their fair share of tax and are, therefore, being subsidized by others who do pay tax in higher tax jurisdictions.

To be fair, some of the companies that take advantage of the present laws point to the fact that they are doing just that—following the law. Notably, in talking about the US corporate tax rate, Jeff Bezos, the founder and CEO of Amazon, has publicly stated that Amazon is "supportive of a rise in the corporate tax rate"[40], and Amazon News responded to Senator Elizabeth Warren's criticism of Amazon's tax payments by saying, "You make the tax laws @SenWarren; we just follow them. If you don't like the laws you've created, by all means, change them."[41]

Supporters of a global minimum tax for multinational companies include The Institute for Public Policy Research Centre for Economic Justice, US Treasury Secretary Janet Yellen, and finance ministers from France and Germany, who both have indicated that they are willing to support a rate of 21%[42]. But the proposal has its detractors, including the CEO of the US Chamber of Commerce[43] and representatives from various low tax countries, including Ireland, where the corporate tax rate is an anorexic 12.5% and being a location that many multinational companies, particularly ones in the digital space, prefer for that reason.

Those in favour argue that a minimum global corporate tax rate would prevent multinational companies from siphoning profits off to low tax jurisdictions and this would put a stop to the unfair advantage they have over companies that can't do that. Some proponents are in favour for ideological reasons, thinking that more money for government and less money for big companies and the powerful elites that control them is a good thing and that this aligns with the goals of the Occupy movement against economic inequality. Some argue that having a global tax rate consistently applied across all multinational companies would stabilize workforce and operations because it would keep the companies from relocating if needed to qualify for a lower tax rate. Secretary Yellen believes that not having a global tax rate will result in a "race to the bottom", but that with the tax in place it will "… make sure the global economy thrives based on a more level playing field in the taxation of multinational corporations, and spurs innovation, growth, and prosperity."[44]

Arguments against….

Those against the idea of a global tax claim that having a global tax would undermine national sovereignty, but the counterargument to this is that shifting profits to offshore tax havens potentially undermines sovereignty to an even greater degree.

Detractors also claim that the proposal is doomed to fail and that it could work against the US and its allies by empowering countries, including China and Russia, that might wish to undermine the liberal national order[45]. There also is the possibility that some countries will opt not to participate in a global tax regime because not being part of it will make them more attractive to multinational companies, so they will derive an even greater benefit.

Critics also argue that setting a global tax rate for multinationals will hurt developing countries, because they often attract the much-needed foreign direct investment by creating tax disparities that the multinationals can exploit and create local jobs in the process. Without lower taxes, multinationals may no longer be interested to invest in places like Moldova with its 12% company tax rate or Uzbekistan where the rate is even lower at 7.5%, and the economies in those countries and others like them will suffer as a result. Further, the critics suggest, the multinationals themselves will simply pass on any additional tax that they must pay to consumers or to suppliers, making their goods and services more expensive, or squeezing smaller companies that perhaps can't afford to be squeezed.

Several of the larger developing nations have sponsored a rival tax proposal through the UN that focuses on taxing digital services companies, this move seemingly being motivated by bad feelings about the paltry taxes that some of the largest US companies (including the FAANG gang) pay in many countries where they make large profits. The plan would see digital companies being taxed on revenues in the place where they are generated rather than where the company is headquartered.

New tax rates

Presently, the worldwide average corporate income tax rate across 177 jurisdictions is 23.85%, the average rate for the G7 is 24% and the average rate for Africa is 28.50%[46]. Noting this, we propose that a global minimum tax rate of not less than 23% be set for all companies, regardless of size or location. This is less than the present global average, higher than the present 21% in the US, just under the average rate for G7 nations and substantially lower than the rate for Africa. If the tax is 23% across the board then there will be no incentive for tax location arbitrage by companies. Arguably, bigger and more profitable companies—say, the top 100 multinational companies—can afford to pay more and should pay more. Setting a higher rate of 26% for them would mean that they contribute more and removing some of the criticism that has been directed at them.

Under this new tax arrangement, we suggest that 2% of the tax paid by the top 100 multinational companies and 1% of all tax paid by every other company go into a fund administered by the UN[47] for distribution to less developed nations, directing it specifically towards infrastructure improvements in developing nations and toward the training and development of their workforce so that they can be elevated to participate more fully in the KIC economy. Additional incentives might also be given to companies to locate part of their operations in less developed countries, thereby creating jobs in those places.

The taxes allocated to less developed nations under this scheme is a relatively small percentage of the total tax that would be generated, but in aggregate the tax dollars would make a huge difference over time by giving a helping hand to lift supported nations up the economic ladder, and it would be a benefit to all by contributing to global peace, stability and prosperity.

> *"Unless we act now, the world is susceptible to the emergence of a deepening global divergence between rich and poor countries. ... The result would be a deeper and longer-lasting crisis, with mounting problems of indebtedness, more entrenched poverty, and growing inequality."*[48]
> —US Treasury Secretary Janet Yellen, in a statement to the Chicago Council on Global Affairs, April 2021

Some critics claim that the economic theory behind a global minimum company tax rate for multinational companies isn't wrong but that in practice in the real world it just won't work. That will most assuredly be true if it isn't tried. Our position is that there are always naysayers and possible negatives and problems, but that if a global compact is achieved then any problems can be regulated. The fact that this hasn't previously happened doesn't mean that it can't happen, and it certainly doesn't mean that we should stop trying.

TAXING AI AND ROBOTS

In the digital and KIC economy where for many tasks human labour has been replaced by robots and AI, will the tax base erode to the point where essential services are not properly funded? It's an interesting question in which many assumptions and possibilities are wrapped. We need to unbundle the problem to provide a sensible answer.

First, consider that we can choose whether people will be displaced by AI and robots. With long-term transitional planning we can retrain or shift workers into new occupations. We expect that people will continue to work; they will just be doing different types of work. Of course, a systemic shift on training and access to education is a fundamental requirement for an orderly transition.

Secondly, in line with greater corporate social responsibility, big tech that is deploying AI that is destroying jobs will be asked by policymakers, shareholders and customers alike to mitigate that damage. We see the biggest tech players as willing participants in massive job creation and retraining programs as they disrupt industry and classic job roles. If not, these companies can expect to become social pariahs where people reject their products and services as a result. Ultimately this sort of corporate culture shift will be the only way major FAANG/BATX players get to manage their brand health in a more socially connected global workplace.

Thirdly, if these corporations don't proactively respond to the employment changes they are thrusting on to the world, governments will end up having to tax the robots, algos and AI they deploy. Corporations will either willingly mitigate the jobs they are destroying or they will be forced to give up their massive pools of cash for the benefit of the communities and people they serve.

The other thing to bear in mind is that the companies that own the robots and use AI will be massively more profitable than they would be without AI and automation. This wealth will extend to filling the public purse through the global taxation scheme proposed above. We could even create regionally-based climate mitigation and smart infrastructure roles that sit above multiple economies funded by this enormous windfall generated by AI.

The question once again comes back to equitable distribution of the wealth. If the largest companies in the world are tech giants obliterating human process-oriented workers and at the same time pooling trillions in profits for shareholders and corporate wealth, we can expect society to legislate that inequity out of the system.

Thus, AI means the overall tax base grows but as the economy advances further and becomes more digital and more dependent on automation, the growth in profits attributable to work performed as a result of automation will grow disproportionately to the growth in profits generated by human labour. As this occurs, several tax considerations arise.

Inarguably, the tax dollars generated must be sufficient to fund social goods. In today's economy this means funding education, healthcare, pensions, infrastructure, defense, trade support and community projects. Additionally, we will need to fund the fight against climate change, future pandemics and other unforeseen challenges that we know are likely to happen. Many more tax dollars will be required to satisfy future needs and expectations than is presently the case. As greater profits will be generated by automation, we will need to ensure that either the taxes paid by the companies that benefit from automation covers that bill, or we will need to move the tax burden from labour to capital by taxing the robots and AI themselves.

A simple approach for existing companies making new investments in technology would be to use pre-investment profit as a single year data point or as an average over several years as a baseline. Investments made in robots and technology would first be fully deductible against these profits so that no tax is paid until the capital cost is recouped. This incentivizes the business to make the investment and allows for a transition period during which the technology is integrated into the business while it operates to produce profits.

Once this is achieved and the investment costs have been fully covered, the company would begin paying tax on an imputed value for the labour substituting input that the robots and AI represent. One way to calculate this would be to scale the tax paid by each company with the portion of wages previously paid to human workers.

For example, assume that the company tax rate is 23% in line with our recommendation above, and that a company had been earning a profit of $100 after total labour costs of $70 had been deducted from gross revenue of $200; the scale factor would be 70/200 = 35% of revenue that is then taxed at what would have been the average rate of tax paid by the workers employed by the company. If that average rate was, say 20%, then the tax would be 20% of the $70, totalling $14.

The $14 would be paid as additional tax by the company, a notional tax on the contributions made by the robots and AI that the company uses. The company would still be ahead by the $56 (being $70 minus $14) that it would have otherwise paid in wages to human employees, because the robots don't get paid and their cost has already been fully recouped by the company to encourage the investment in the first place. The tax paid by the company would be the $14 imputed "robot tax" plus the company tax of 23% levied on $100 profit being $23, adding to a total company tax of $37. We believe that companies should not be allowed to just destroy old-economy jobs and pass 100% of the resultant economic gain to shareholders alone. We need socially-conscious capitalism.

PROPOSAL: TWO-TIER TAX AND DEBT FORGIVENESS FOR CLIMATE ACTION

A growing chorus of scientists, political leaders, economists and people in general are calling for more to be done to fight climate change. Janet Yellen, United States Secretary of the Treasury, has stated that climate change is "the biggest long-term threat the world faces"[49]. So what mechanisms might enable governments to do more?

The 2016 Paris Agreement provides, for the first time, a legally binding international treaty on climate change. Its goal is to limit global warming in the hope of achieving a climate-neutral world by mid-century[50]. The Paris Agreement makes it clear that developed nations should provide financial assistance to developing nations in fighting climate change. For the agreement to work, each nation will need to submit non-binding national plans (Nationally Determined Contributions or "NDCs") specifying the actions that they will take to contribute to the global targets set out in the Paris Agreement.

The NDCs under the Paris Agreement need to be funded. For many nations this will be a big challenge. India, for example, estimates that it needs at least $2.5 trillion to meet its climate change commitments between 2015 and 2030, stating that India has to date mainly self-funded its climate actions, but to scale up those actions will require greater resources[51]. India isn't the only nation that is likely to need money from others to support its efforts in combating climate change. Most developing nations will need help.

According to the definition used by the IMF, presently there are 152 developing countries. These countries account for approximately 85% of the world's population[52]. The debt owed by these countries has more than doubled in the past decade and more than 50 countries have an extreme inability to repay the debt that they owe, and the COVID-19 crisis is making the problem worse[53].

Aggregate debt commitments on external debt owed by emerging markets and developing countries for the year 2020 was around $11 trillion[54]. In April 2020, G20 nations agreed to a "debt service standstill" until the end of 2020[55] to support the poorest nations. This is a small and temporary relief[56]. It has been claimed that this debt poses a global development emergency, much like the COVID-19 pandemic has created a global health emergency[57].

Many developing countries are growing at a faster rate than many developed nations. At some point these developing countries will catch up and be classified as developed nations and be treated as such. However, some of the poorest and least developed nations will need the most help.

Considering the economic disparities between developed, developing and least developed nations, capacity to pay, and ability to presently contribute to funding the fight against global existential threats, we propose that a climate change action fund be instituted with contributions being calculated as 2% of GDP for developed nations, 1% of GDP for developing nations, and no financial contribution from the world's least developed nations. This is effectively a climate tax in two tiers to be paid by qualifying nations.

Once a developing nation is categorized as developed by the IMF, it then pays 2% of its GDP into the fund. Nations that do not meet their obligations could be sanctioned by those that do with trade embargoes, for instance. Provided that the negative impact to GDP for the non-complying nation of any sanction imposed is greater than its climate fund obligation then the rational behaviour is to comply and pay the 1% or 2% into the fund. Besides, complying nations will keep their "good global citizen" rating with other nations. That will matter more in the future as people from all nations come to view climate change as a global problem to be solved by everyone. Freeloaders will not be viewed kindly by those from outside, nor by those from within.

For the least developed nations we propose that the Debt Service Suspension Initiative (DSSI) by G20 nations translate into full debt forgiveness of all public sector debt owed to them by the least developed nations. The debts that are forgiven will be credited against the incremental contributions[58] to the climate action fund that would otherwise need to be made by any G20 nation, thereby offsetting the 2%

tax on GDP. Once the debt has been fully offset, the 2% tax would apply in full.

Other countries outside of the G20 would be encouraged to participate. Perhaps those nations might do so by forgiving, say, half of the debt that is owed to them by the least developed nations. This would create the feeling that when it comes to fighting climate change, we are all in this together.

We can fight climate change and prevent foreseeable future tragedy, but we must make some sacrifices for that to happen. Sure, 2% of GDP is a big tax. But climate change is a big problem. Ask someone who wants to live but who has been diagnosed with a terminal illness what they would give to be cured and a common answer comes back: "anything". Faced with crisis, they would pay anything to fix it. The planet is facing a crisis. We need to pay what it takes to fix it.

Critical: Advanced nations must implement UBI

UBI has been debated and discussed and trialled for a long while. We discussed some of the successful trials in earlier chapters. Overwhelmingly, the evidence shows that UBI works as intended and doesn't create the 'couch potatoes' that conservative politicians claim. In fact, the opposite appears true—that UBI stimulates work and activities that matter, that people are passionate about. UBI planning is needed now because the changes in AI and automation that are coming will happen too quickly to mitigate in real-time. Productivity in certain industries is going to boom, but it will collapse in others.

We suggest that developed countries should work on implementing a UBI for their citizens. With UBI in place, a giant step toward fuller economic inclusiveness will have been taken and we will be better positioned to transition to the digital KIC economy. Once UBI is operating, it can be adjusted as needed to close gaps, to make it more effective and to deal with any unintended consequences. At some point, the economy may evolve to the point where a UBI is no longer needed. It may be transitory to get us over the hump of present and foreseeable challenges—only time will tell.

Opposition to UBI is because most people are focused on what they have now and their fear of losing it, rather than on what could be and should be, and what is best for humanity overall. We need longer-term economic thinking and planning. We need this at the government level and at the people level. We need to give people time to think and plan and reflect and learn and acquire new skills. UBI provides this time. UBI won't create income equality, but it will create a more level playing field on which there is greater equality of outcome. UBI will help to ensure a more equal access to opportunity. Individuals will still be responsible for making their own choices, and those who are smartest or most talented and who work hardest will still enjoy higher levels of success and reward than the rest; but poverty will be eliminated, inequality diminished, and no one should feel like they are a social outlier.

Taking sabbaticals

To encourage workers to work until later in life, and in recognition of the fact that in the future economy many people will have two or three careers (or more) in their lifetime, we propose that UBI-sponsored sabbaticals be introduced along the way. This could take the form of a year off after each 15 years of full-time work on full pay, with benefits to go back to school and learn new skills and explore new ideas. A refresher, a capability builder, and a productivity booster (via the new ideas) all in one.

The year-long sabbatical could be kept flexible so that people on sabbatical have time to develop interests and pursuits outside of work, and to spend time with family and friends and deepen their relationships in their community. The retraining at older ages might be focused on teaching people how to mentor others and how to teach others the core soft skills that they have developed during their careers, or how they can contribute to the community in their twilight years.

By the way: we expect that at the present rate of advancement in medicine, health sciences and lifestyle changes, by the year 2040, being 77 years old will be considered far too young to retire and do nothing. The idea that being "old" equates to being less useful and more reliant on others will become a thing of the past as health, mobility and cognition improve through health-tech and gene-therapy advances.

Zero sum doesn't work

For the future economy to work in a socially harmonious way it cannot be zero sum. The *Human species* needs to win. Not everyone needs to win in the same way, but people cannot feel that they have been excluded, side-lined, overlooked and disadvantaged. The potential for this to happen to more and more people as AI and climate change impact our planet is massively acute. We either collaborate or we fall apart at the seams. The majority of the world can't feel like others are winning while they are losing. In fact, no one should feel that way, but we have a long way to go to restore and ensure a sense of fairness and equality that will allow for humanity to grow beyond tribalism and into shared prosperity.

Equality of opportunity is a must. Same effort and same results must be equally rewarded, both in terms of pay and recognition. It won't happen in the near future, but long-term this will have to be true regardless of where someone lives. Globalization will mean that it must be so. New communications technologies like global internet satellite constellations, international health-care research, automated shipping networks and such, are an accelerating force in this regard. The 21st century digital economy with its AI, robotics, alogorithms, new decentralized financial systems, jumps in longevity, the design of future smart cities, new population dynamics, new models for education and more, mean we must increasingly act in concert as markets and as nations.

The rise of Asian economies will mean that long-held sacred ideologies in Western societies will need to change. This is a reality that will at some point be accepted. Resistance from "advanced" economies that believe this new reality undermines their authority could create global economic conflict that hurts everyone. The US might even feel that using their military to combat the rise of China might be their only remaining option as China's GDP and trade obliterates US advantages. The more the US doubles down on traditional industry, infrastructure and policy, the more apparent the gap between China and the US will be. This is the biggest argument for a bottom-up revamp of the US economy to a KIC platform.

The biggest economies of the 21st century, such as the United States, China, India and other developed nations, have a responsibility to not only let others in but to create an environment in which equality is prized. That humanity's future is safe, sustainable prosperity is assured, and environmental protection is guaranteed, must be a doctrine that we all embrace and argue for.

Collectively, we need to future-proof the economy. Change and uncertainty tends to foment rivalries, jealousies and fears among people who aren't communicating, who don't try to understand one another, and who think that their differences set them apart—that there must be an "us" versus "them". Never was this more apparent than during the COVID-19 pandemic, and it didn't help a single citizen cope economically or improve their health. If we continue to play this game, there is no happy utopian future, the goal itself remains unrealistic and unattainable. You might believe that a planned, abundant future is impossible. We pray that it isn't.

Regardless, in striving to work together, in putting aside the rivalries, jealousies and fears, and in viewing differences as a source of strength that helps to advance humanity, we have a much better chance of surviving the economic uncertainty and chaotic change that is coming.

Endnotes

1 Certainly this applies to the GFC.

2 Although the UBI trials mentioned in previous chapters indicate that the opposite may also be true.

3 Typically through a central bank.

4 https://www.bennettinstitute.cam.ac.uk/research/research-projects/wealth-economy-social-and-natural-capital/.

5 Ever had an important test to study for and spent time cleaning out the closet instead? Substituting one non-urgent but easy task for another more urgent critical task. Feeling productive and assuaging the guilt of laziness, while missing the target. That's how we define constructive apathy.

6 We subscribe to the credo that big ideas means taking risk, that combined with forward thinking and hard work would lead one closer to success than otherwise might be achieved. A favourite teacher put it this way: "Boy, aim for the stars and you might just hit the moon, aim for the back fence and you'll end up dribbling on your shoes. There's greater disappointment in ruining your shoes than in hitting the moon."

7 United Nations Development Programme Mohammed Bin Rashid Al Maktoum Knowledge Foundation, Global Knowledge Index 2020, p.8.

8 Sources: United Nations Development Programme Mohammed Bin Rashid Al Maktoum Knowledge Foundation, Global Knowledge Index 2020; United Nations, Statistics Division; Statistics Bureaus and official estimates for population for various nations.

9 2021 Internet Accessibility Index, Broadband Choices.

10 Note that Singapore is not included in the 2021 Internet Accessibility Index, but made public its plans to become a leading digital economy and is among the leaders in making progress toward that goal.

11 https://www.investindia.gov.in/sector/it-bpm.

12 Human Capital flight or the "brain drain" effect has been observed for many years in economies like Australia, India, and even China where the best people leave for offshore opportunities because the salaries are magnitudes higher than at home: see https://en.wikipedia.org/wiki/Human_capital_flight.

13 Peter Vanham, "A brief history of globalization", World Economic Forum, 17 January 2019.

14 It is noteworthy that President Xi in Davos in January 2017 stated that China "came to the conclusion that integration into the global economy is a historical trend … is the big ocean that you cannot escape from" and called for a more inclusive globalization.

15 Philip Meissner and Christian Poensgen, "Which countries are making the most progress in digital competitiveness?", World Economic Forum, 7 September 2020.

16 For a detailed discussion on Smart Cities and how they will evolve and what services they will deploy, see: *Augmented: Life in the Smart Lane*, Chapter 11, "Augmented Cities with Smart Citizens".

17 *Financial Times*, "Standard Chartered Acused of Hypocrisy over Climate Change", Attracta Mooney and Stephen Morris, May 2021.

18 CNBC, "Why a 'crushing' day for Big Oil represents a watershed moment in the climate battle", Sam Meredith, 27 May 2021.

19 http://documents1.worldbank.org/curated/en/258321468327925026/pdf/339460trn126120EENote2.pdf.

20 https://pressroom.ifc.org/all/pages/PressDetail.aspx?ID=18568.

21 https://www2.deloitte.com/content/dam/Deloitte/global/Documents/About-Deloitte/deloitte-2020-millennial-survey.pdf.

22 We hope they're wrong.

23 Danielle Dhillon, MBA student, Haas School of Business, University of California, Berkeley as cited at https://www.ifc.org/wps/wcm/connect/news_ext_content/ifc_external_corporate_site/news+and+events/news/insights/impact-investing-for-growth.

24 https://thegiin.org/impact-investing/need-to-know/#how-do-impact-investments-perform-financially.

25 President Joe Biden, remarks during first address to US Congress, 28 April 2021.

26 Aldous Huxley, *Brave New World*, New York: Perennial Library, 1946.

27 The motto of Huxley's World State is "Community, Identity, Stability".

28 J. Lee Thompson, *Theodore Roosevelt Abroad, Nature, Empire, and the Journey of an American President*, Palgrave Macmillan, 2010.

29 The only independent nation states not in the UN are Vatican City and Palestine. In the case of Palestine, it was granted non-member observer state status in 2012.

30 Almost too awful to contemplate, another world war is a sure-fire way to retard humanity. The destruction of war will surely fast forward us to THE END. Destruction in any future war would not be limited to death and loss of buildings but we'd also suffer forever the environmental catastrophe that modern weaponry would cause. There would be no winners.

31 https://www.un.org/sustainabledevelopment/sustainable-development-goals/.

32 https://www.un.org/en/about-us/un-charter/chapter-5.

33 https://www.theguardian.com/world/2015/sep/07/what-has-the-un-achieved-united-nations.

34 There are notable exceptions.

35 *Regulatory Impact Analysis: A Tool for Policy Coherence*, OECD, 11 September 2009.

36 World Economic Forum Global Competitiveness Index, Burden of Government Regulation, 2017.

37 Different tax rates have been proposed, with the US treasury viewing 15% as a floor, with hopes that the final rate will be higher.

38 The OECD has been highlighting tax avoidance by multinational corporations since 2009 and has been recommending changes to modernize the international tax architecture.

39 https://www.oecd.org/g20/topics/international-taxation/.

40 https://www.aboutamazon.com/news/policy-news-views/a-message-from-jeff-bezos.

41 Amazon News response to Senator Elizabeth Warren on Twitter, 26 March 2021.

42 https://www.nytimes.com/2021/05/20/business/economy/global-minimum-tax-corporations.html.

43   https://www.nytimes.com/2021/05/20/business/economy/global-minimum-tax-corporations.html.

44   https://home.treasury.gov/news/press-releases/jy0101.

45   https://foreignpolicy.com/2021/04/12/global-minimum-corporate-tax-janet-yellen-bad-idea/.

46   Elke Ansen, Corporate Tax Rates Around the World, 2020, The Tax Foundation, 9 December 2020.

47   Or a purpose-created organ with similar authority and remit to improve lives and transform communities.

48   https://home.treasury.gov/news/press-releases/jy0101.

49   https://home.treasury.gov/news/press-releases/jy0101.

50   https://unfccc.int/process-and-meetings/the-paris-agreement/the-paris-agreement.

51   Preliminary estimate at 2014–2015 prices: see "India's Intended Nationally Determined Contribution: Working Towards Climate Justice", submitted to the UNFCC, 1 October 2015.

52   https://www.worlddata.info/developing-countries.php.

53   https://jubileedebt.org.uk/press-release/poor-country-debt-payments-soar-to-highest-level-since-2001.

54   https://www.brookings.edu/blog/future-development/2020/04/13/what-to-do-about-the-coming-debt-crisis-in-developing-countries/.

55   This Debt Service Suspension Initiative (DSSI) has since been extended through to the end of December 2021.

56   https://www.oecd.org/coronavirus/policy-responses/a-debt-standstill-for-the-poorest-countries-how-much-is-at-stake-462eabd8/.

57   https://www.brookings.edu/blog/future-development/2020/04/13/what-to-do-about-the-coming-debt-crisis-in-developing-countries/.

58   Though not against the baseline commitments to the climate action fund, which must in any case be paid.

CHAPTER 10

THE RISE OF TECHNOSOCIALISM

Humanity is at an inflection point. We're only going to get one shot at this. The issues that we cover in this book will, and must, change us fundamentally as a species. Driving these changes are the most fundamental questions that we will probably ever grapple with. We either double down on a system that has prioritized the chasing of profit as an organizing principle, or we prioritize the prosperity, survivability, health and happiness of the species. It is not a hard decision when you frame it like that, but it is an incredibly complex one to sell when the whole planet is oriented towards economic competition, engineered scarcity, and believing that your individual rights or rights as a nation trump the rights of the rest of your fellow humans. This is modern tribalism pitting us against everyone else based on an age-old but arbitrary classification of wealth and advantage.

Humanity has always been at its most powerful, most enlightened, and effective when it collaborates. But capitalism, nationalism, religion, and other contrived structures have divided people in ways that continue to be counterproductive to the broader aims of the species.

If you have gotten this far, it's fairly likely you're not an all-out climate change denier, but you may still have doubts that humanity is responsible for the changes in the climate that we are observing. That's fine, the reasons why the climate is changing are largely immaterial to our future. We must adapt regardless of causality.

Global temperature rise is easily measurable—we know that over the last 150 years it has accelerated, and we can anticipate that by the end of this century temperatures will rise from 1.5–3°C. We know that glaciers are melting at the fastest rate observed in hundreds of years. This is not like ice floating in a glass—the classic retort—this is ice on top of a landmass that when melted will be dumped into the sea in trillions of litres annually, accelerating sea rise 14,000 tons of water per second alone just from melting Arctic ice. Thus, we know that the rate of sea rise is predictable within ranges and this allows us to identify which cities will be the first to be inundated and overwhelmed.

Extreme weather events are accelerating—100-year floods, 100-year hurricanes, 100-year droughts, leading to 100-year wildfires or bushfires that are now commonplace each season. The cost in human lives and the economic impact of these events grow each year. We keep kicking this can down the road, because it's an almost unimaginably large problem to grasp in respect to reversing these trends, and yet some keep insisting this is just normal weather cycles. At what point do we accept that the annual cost of these "acts of god" outweigh long-term resilience and mitigation action? When the global insurance industry collapses? Because it will when it can no longer underwrite those annual costs. Then it will already be too late— we will be on a wartime footing against the climate, and perhaps against each other.

"In 2019, weather-related hazards triggered some 24.9 million displacements in 140 countries."
—UNHCR figures, 30 November 2020[1]

In China, around 20–30% of the population will be displaced. Bangladesh and India will be hit hard, with upwards of 200 million people displaced between now and 2050. Vietnam, Indonesia, Thailand, the Philippines and Japan will also be affected. For nations like Tonga and Maldives, sea rise over the next 30 years is going to be absolutely devastating, affecting 80% of their land area. Miami and New York will be forever reshaped, with large parts of the cities unlivable and flooded permanently. In Europe, the Netherlands, Italy, Greece and London will be among the places most dramatically impacted.

In Africa, at least 20 cities, including Cairo, Cape Town, Lagos, and Kinshasa, will be particularly vulnerable to rising seas. At the top of the list is Lagos in Nigeria, where half of the expected 33 million inhabitants in 2050 are likely to be affected.

The trajectory of these events is already baked in—we have simply not done enough historically to prevent these events from now happening. Instead, we will have to adapt, and we will need to do so quickly. In the space of just three decades. This is not "climate change", it is a global climate catastrophe affecting every person and nation on the planet.

In the midst of all of this, we'll also be dealing with the emergence of a competing intelligence infrastructure on the planet—the emergence of Artificial Intelligence (AI). The initial effects of AI will be simple. Mass automation of society, leading to mass unemployment that reframes our connection with work, resource utilization, commerce and capitalism itself. More efficient government, more efficient resource utilization and drivers like longevity and the generation of massive wealth will incentivize the market like no other boom in humanity's history.

The emergence of AI and the effects of the climate catastrophe on the planet are epic changes the likes of which humanity has never seen before, and we will live through both simultaneously. Both give us incredible incentives to work together, to reshape our society from one that is largely reactive and short-term focused because of present economic incentives, to a long-term commitment to the planet and the species. Either that, or we continue to argue over GDP growth and who's economy is more powerful, which form of politics and economics is better, and whether or not racism, nationalism, and discrimination are justifiable, until society inevitably collapses.

We can yet choose better outcomes

There is cause for incredible, unbounded optimism, however.

Humanity has never before had the power to reshape the planet as we have today. Over the next century we will hone these geo-engineering skills to a fine art, making our cities and homes resilient and sustainable. We will relocate close to two billion people by the end of the century to new smart cities away from affected coastlines and into cities that better support sustainable living. We will use technology to extract CO_2 from the air, initially sequestering that carbon and then turning it into exotic materials like carbon nanotubes. We will build sea wall defences and modernize infrastructure. We will work methodically to extract pollutants from the sea, moving away from massive fishing industries to sustainable fish populations and lab-grown alternatives.

We will automate farming and re-engineer the food supply chain, distributing food production between cities and farmland using autonomous systems. We will move away from most of the cattle and poultry farming we have today to lab-grown proteins. We will use vertical hydroponic farms in our cities that use fractional amounts of water used by the old methods. We will eat healthier and do so with a much smaller carbon footprint than was used by the previous century's farming practices.

Today, 50% of global commodity trades are energy-based, but over the next decade we will evolve to a global, clean, extremely low-cost energy infrastructure based on the much lower cost of renewables. We will have autonomous transportation. We will have intelligent healthcare, transport, governance, emergency services, recycling, and much, much more.

Artificial Intelligence will release us from the burdens of day-to-day economic servitude, changing, for most of us, the role of work in our life to being something that we're passionate about, not something we need to do to survive. We won't have people going hungry while working 70 hours a week on minimum wage.

AI will give us the tools to live longer and healthier lives. It will help us answer some of the biggest questions of the universe, unlocking incredible computing power and advances that make what was once thought science fiction a daily reality.

Automating government—big government made small

The biggest changes in society, however, might be the unintended consequences of simply using AI to automate government. Remember that when we started this journey in Chapter 1, the promise of AI to radically reform big government was the hallmark of redefining some of the negative attitudes towards socialism. When you can make big government tiny through automation, you not only get a bigger bang for your buck out of tax revenues, you get to reinvest that in the health and education of citizens, along with broader services and infrastructure.

By the mid 2030s, regulations will be encoded so that AI can decide on whether or not you or an organization is in infringement of the law. This will lead to AI-based sentencing designed to eliminate political, racial, gender, or religious bias. AI-based contract law will interpret and decide on judgements based on said laws, precedents and the contracts themselves. Image recognition, cameras, and sensors embedded in the world around us, will allow us to identify criminals faster. Crime will be reduced both as a result of better policing, but also due to UBI reducing want. The cost of compliance and law enforcement will plummet, jails will be emptied, and a citizen won't need to fear the police for non-compliance. (However, it should also be noted that this same technology could be used to reinforce existing biases and to suppress dissent.)

There are a few economies already using automation of government and city services effectively, demonstrating the advantages and traps of Artificial Intelligence. We'd like to share three examples that illustrate what is possible, and why clear policy decisions are critical early on.

Facial recognition in China

China has employed vast facial recognition technology, which is being used primarily in financial services, but is expanding into other areas. Children in school throughout China use facial recognition to enter school campuses, which could broadly be seen as a positive security measure. You can pay for goods and services at a store using your face, instead of using cash or plastic cards. During the pandemic, this system was used to create completely contact-free transaction capabilities, although the image below is from before these updates.

Figure 1: Facial recognition has seen broad use in China for payments, dramatically reducing fraud rates. (image: Ant Group)

The Ant Group annual report[2] shows that Alipay has used biometrics, including embedded fingerprint and facial recognition technology at the handset level, and facial recognition at the point-of-sale, to dramatically reduce identity theft and fraud rates across China. In 2020 during "Single's Day" (the equivalent of Black Friday or Cyber Monday, but initiated to celebrate all the 'single' people in China), Alibaba and JD.com racked up $115 billion in sales, 70% of which was done using biometric payments technology.

On the flip side, in 2018 the *People's Daily* newspaper claimed on Twitter that China's facial recognition system could scan the faces of all 1.4 billion citizens in just one second. In December 2017, BBC TV reporter John Sudworth travelled to China and was given access to China's national facial recognition capability. The Chinese authorities demonstrated that they were able to locate Sudworth "hiding out" in a remote Chinese village within just seven minutes[3], even though he had never registered with the system in the way locals do. The West has repeatedly argued that facial recognition has the potential to be abused by governments, but right now China is simply pioneering digital identity infrastructure in respect to broad digital services access. We agree, longer term, that more debate, regulation and policy setting in respect to AI and technology like facial recognition is critical to avoid misuse.

The fact is that China's system must become a template for the rest of the world when it comes to digital identity infrastructure. The 21ˢᵗ century will provide most of our day-to-day services through a digital layer, thus we need to do much, much better than your signature, date of birth, an address (that can quickly become obsolete), your mother's maiden name and so forth to secure identity as those details are no longer secure. We need core identity infrastructure that allows access to the most advanced services available—whether to educate our children, rent an apartment, get healthcare or transact with our cyber dollars. Identity reform is a prerequisite for 21ˢᵗ century capable KIC economies.

Shenzhen China—A Smart City that demonstrates the way

Shenzhen is perhaps the most automated and smart city on the planet. While other cities like Amsterdam, Tokyo and even New York use some automation (AI-based registration tag/number plate reading for road tolls, autonomous traffic algorithms, etc.), Shenzhen has taken this to a whole new level with its "Intelligent Twin" city-wide AI system, moving from e-government to smart-government.

Shenzhen has roughly 3.5 million automobiles, compared with the roughly two million that inhabit New York City. This makes Shenzhen the city with the highest density of motor vehicles in China. To better manage this growing vehicle population, the Shenzhen Traffic Police implemented AI and 5G/Edge computing to tackle traffic flows, law enforcement, congestion reduction, and basic control and command. This has reduced wait time at intersections by nearly 20% and increased traffic capacity by an additional 10%.

If an accident occurs on a Shenzhen motorway, image recognition will automatically determine if a police patrol vehicle is required, if an ambulance should be dispatched or fire and other emergency services vehicles. This response means that vehicles are moving well before police attend a scene. Ambulances and fire services vehicles are given green lights all the way to the scene of the accident via an autonomous system. In a study for the US National Institute of Health, it was found that Shenzhen, with its 111 ambulance stations and more than 500 ambulances, consistently outperformed US 911 response times and EMS cost effectiveness by significant margins[4].

Today in Shenzhen you don't get stopped by the police for speeding, for not wearing a seat belt or if you have a tail light out. You get issued a fine that is sent to your mobile phone. For many first offences you just get a warning, although the system has already demonstrated its capabilities if you continue to abuse traffic laws. You might argue that this is an infringement on your basic civil rights, but if you live in the US and you are an African American, this system would be a whole lot better than getting stopped by a patrol vehicle and potentially being harassed. There's arguably no reason why we need physical intercepts for traffic stops, given the capabilities of image recognition today. It would massively reduce the cost, accuracy and fairness of policing. Ultimately, Autonomous Vehicles will totally disrupt motor vehicle infringement policing and revenues in any case.

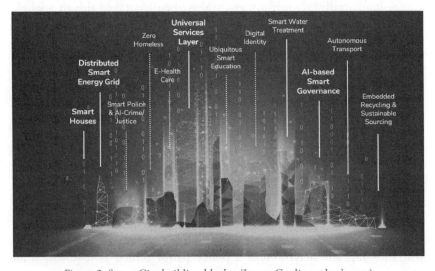

Figure 2: Smart City building blocks. (Image Credit: author's own)

Taiwan's AI policy tech

Taiwan has a 30-something whiz kid who runs the digital platform for the current government administration. Audrey Tang is a Taiwanese free software programmer who has been described as one of Taiwan's ten greatest computing personalities and was the first non-binary gendered official to join Tsai Ing-wen's cabinet. Tang's first initiative was to introduce the g0v project (purposely replacing the 'o' with a zero). Secondly, they crowdsourced a national hackathon to mobilize against the COVID-19 pandemic, specifically solving the problem of mask distribution using tech like Google Maps. Thirdly, Tang created a policy debate platform as part of the g0v platform that uses AI and social media in an attempt to reach consensus on issues that previously divided Taiwanese constituents.

"Whether the challenge is fighting disinformation campaigns orchestrated by hostile powers or the existential threat of a virus run amok or simply figuring out how to regulate Uber, Taiwan is demonstrating the best ways technology can be used to marry the energy and talents of civil society with the administrative powers of government bureaucracy."
—How Taiwan's Unlikely Digital Minister Hacked the Pandemic, *Wired*, 23 July 2020

This platform is part of what is now known as vTaiwan, or "virtual Taiwan". The achievements of the system to date include: a crowdsourced bill successfully passed through parliament on closely held company law; the resolution of a disagreement between civil society activists on the topic of internet alcohol sales; and the ratification of several items on ridesharing (Uber) regulations.

These are fairly small policy positions in the scheme of things, but the platform was able to successfully inform constituents and allowed sensible debate that resulted in broad consensus. The issues were addressed by using deep learning, data mining and behavioural mapping to identify areas where consensus might be possible. This allowed vTaiwan to explore areas that politicians may have avoided, but where AI determined there were positions and arguments that could bring people together.

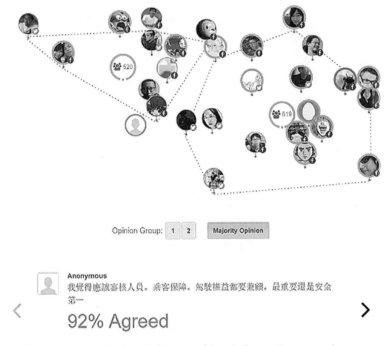

Figure 3: Screenshot from Pol.is, one of the platforms vTaiwan used to debate ridesharing services. (Source: vTaiwan)

vTaiwan is a promising experiment in participatory governance. The process was designed to facilitate constructive conversation and consensus-building between diverse opinion groups. It achieves this by creating several stages, including an initial 'objective' stage for crowdsourcing facts and evidence, and a 'reflective' stage using the mass deliberation tool Pol.is, which encourages the formation of 'rough consensus'. Lastly, key stakeholders are invited to live-streamed, face-to-face meetings to draw up specific recommendations that have subsequently been turned into legislation.

vTaiwan shows that there is a way to use technology to create real-time governance on major issues, and both educate citizens on the issues and reach consensus. This would allow much more participatory government than simply voting your side in. It is a logical evolution of the political landscape, but gets us away from representative government by individual parliamentarians, senators or congressional representatives, and allows a skeleton administration and then cabinet members who are there to execute the will of the people.

Policy issues we need to resolve during the 21st century

Policy drives both changes to laws and changes to government budgets, but we can predict that certain policy decisions will also become inevitable due to macro changes, challenges and developments taking place on the world stage. These are not all issues that we will agree on, meaning that vigorous debate and our collective responses will determine which specific paths humanity follows, including:

Inequality, Discrimination and Racism:
This will continue to be a fight between the effects of economic uncertainty and the continued effects of marginalized capitalism, along with unenlightened views of human dominance or supremacy when it comes to religion, race, gender or class. By the second half of this century, we will have either agreed that a minimal quality of life for all citizens is a mandatory threshold for the global economy, or we will have likely surrendered to permanent class conflict. But the ethics of human existence are about to face the ultimate test. If you think gender identity is a complex issue, wait till we develop transhumanist cybernetically enhanced humans, gene-therapy and transgenics.

Climate Change:
- **Crop Failures, Food, and Agriculture Agility:** Major changes in farming practices, crop suitability, and land use will be an effect of rising temperatures. Already certain places like the wine-growing regions of France are showing significant sensitivity to temperature

increases, resulting in crop losses. We will need to adapt. Lab-grown proteins, vertical urban farming and other strategies, will be required.

- **Sustainability, Reuse, and Recycling:** Big social pressure to impact the environment as minimally as possible. It will become unfashionable and socially abhorrent to use single-use plastics or a smartphone that you throw away annually.

- **Climate Repair:** Extracting plastics from the ocean and carbon from the air, restricting access to habitats of endangered species, widespread bans on fishing, reduction in cattle farming, regreening of the planet (1 trillion trees initiative), etc.

- **Mass Migration and Eco-Refugees:** How we deal with hundreds of millions of climate refugees is potentially the biggest single direct outcome of climate change. Ignoring the problem will make it far worse in terms of human lives.

Artificial Intelligence and Robotics:
- **AGI Sentience:** At some point in the next 20 years a set of AI algorithms will gain the equivalent of human sentience and self-awareness—Artificial General Intelligence. This will require us to figure out where intelligent, sentient non-human entities fit in our society. There will be attempts to isolate and control these intelligences, both of which will fail. How we live with each other as fellow humans in the 21st century will provide a proxy for how we respond to the emergence of AGI. The greater the inequality that remains, the more fractious the emergence of AGI will be.

- **Living with Robots:** Robots will outnumber humans by 2035. We will live with physical robots and software robots (algorithms) across every sphere of human existence. Learning to work with robots and integrating them into our lives will be similar to the effect the telephone, television, and internet had on society.

- **Intelligence Augmentation and Neural Interfaces:** To combat the emergence of AI, humans will seek to augment intelligence, but even before that we will use brain-computer-interfaces (BCI) like Neuralink to not only connect our brains directly to the Edge Computing Cloud, but to overcome disabilities and enhance human performance, integrated with robotics and cybernetics.

- **Transhumanism and Cybernetic Augmentation:** Closely related to Intelligence Augmentation is the augmentation of our human form with bio-enhancements and cybernetics. This will start with things like robotic prosthetics that outperform natural human limbs, in vitro gene-therapy to remove chronic diseases and implants that regulate insulin levels for diabetes sufferers. This will evolve into entirely new fields of augmentation, like enhanced vision and hearing, transgenic organs, body modifications, etc.

- **Artificial Intelligence Regulation:** Due to the increasing role AI is playing in society, we can expect whole new disciplines in law and regulation to spring up regarding the use of AI.

Smart Cities and Infrastructure:

- **Smart Infrastructure:** From energy, water and waste management, to broad automation of transportation, emergency services and policing, expect civil resource management to be heavily optimized by AI.

- **Cheap Clean Renewable Energy Sources:** Smart energy management devoid of fossil fuels, distributed energy grids and grid-level energy storage capability will not only reduce our energy costs and footprint to a fraction of that of the late 20th century, but be critical in preventing much worse climate effects in the second half of the 21st.

- **Climate Resilience:** Due to the effects of climate on infrastructure and cities we will need to make our roads, rail, electrical and water infrastructure all flood, fire and heatwave resistant.
- **Food Supply Chain Management:** Due to the impact on farming and food supply chain issues caused by climate change, and the push for sustainability and low-impact farming practices.
- **Autonomous Transportation:** Putting millions of truck drivers, cab drivers and delivery drivers out of work, autonomous transportation will save a million lives annually, but displace millions of workers.

The Digital Realm:

- **Virtual Worlds:** Competing with our own reality will be alternative realities presented in virtual worlds. The metaverse worlds of Snow Crash and Ready Player One will soon become a reality, providing people with both an escape and virtual careers that compete side by side with the real world.
- **Digital Copies:** Combining AI with Avatars, you'll soon be able to have virtual self answer video calls on your behalf as a proxy. Our avatar proxies will be able to do a pretty good job of approximating our behaviour once we train them appropriately.
- **Digital Resurrection:** We are already seeing dead actors being recreated in CGI for movie roles, and the technology will soon make virtual actors indistinguishable from the real thing.
- **Digital Consciousness Upload:** With advances in neural scanning, neural nets and our understanding of

consciousness, sometime in the 21st century we will see attempts at loading our consciousness into the digital realm. This is an alternative form of longevity that might be cheaper than extending our biological lifespans.

SPACE EXPLORATION

Space Exploration and Colonization:

- **Space Tourism:** The cost of space tourism is going to rapidly decline from the current price point of many millions of dollars, to $200–250,000 per flight, to the equivalent of the cost of a first class round the world ticket and lower.
- **Space Manufacturing and Mining:** The space mining industry won't be viable until the late 2030s or 2040s, but will materially change the economics of the planet. It will be necessary for us to stop using resources on earth also. Mining industries won't fade quietly.
- **Permanent Lunar Settlements:** Just like we've done with the International Space Station, we'll start to see permanent lunar settlements over the next decade, which will start another space race between China, Russia, and the US. With the issues of climate change and large-scale unemployment, such pursuits will be heavily criticized as poor public expenditures, as they have been historically.
- **Martian Colonization:** Building cities on Mars and beyond is human destiny, and as Elon Musk likes to point out, it's a critical insurance policy for humanity to escape the possibility of extinction from either another dinosaur-killer comet, or our own self-annihilation.

Healthcare and Longevity:

- **HealthTech Reform:** Reform of the healthcare sector to include genomics, gene-therapy, 3D bio-printing, microfluidics/lab on a chip, AI diagnostics, robotic prosthetics, personalized medicine, etc. along with radical simplification of administrative, supply chain and regulatory systems leading to the elimination of monopolistic insurance and pharma practices will ultimately lead to dramatically cheaper and more effective healthcare for all.

- **Rolling Pandemics:** Coronavirus won't be the last global pandemic we have this century. Permafrost microbes and ancient viruses released from thawing glaciers, the Arctic and Siberian permafrost layers could bring us long-dormant, ancient diseases that our bodies no longer have immunity for.

- **Longevity Treatments:** The ability to extend human lifespans will change the way we view everything from our development years, population growth and stability, our place in the cosmos, to the nature of home, work and legacy. The biggest issue facing this development will be accessibility. If limited to only the wealthiest humans, this will skew human development irrevocably, accentuating class conflict dramatically.

Economics:

- **Workforce Development:** Investments in education and training that need to cover the full age spectrum from early childhood to workers over the age of 55. We also need to incentivize companies to allow time for employee development within work hours.

- **Free Child Care and More Women in the Workforce:** Providing incentives to get more women in the workforce will be critical, even if for part-time roles. As populations shrink, the need to have women employed and participating more extensively in the workforce is critical. A study by S&P Global in 2017 found that if in the US women entered and stayed in the workforce at a rate comparable to that in Norway, then the US economy would be $1.6 trillion larger—that's an increase of more than 8% in GDP.[5] Other studies in other locations have also found that increasing workforce participation by women would have a dramatic positive improvement on GDP.[6] But this requires changes to child care and tax incentives that make this possible.

- **Mature Age Labour Force:** Ageism and bias prevents us from accessing the benefits of an ageing workforce. GDP growth would be achieved by increasing the pension age and by restricting access to pension funds until workers reach older pension ages, even if with part-time work. By the 2040s as longevity improvements become apparent, we suggest as a transitional arrangement extending the mandatory working age by 10 years for anyone presently aged under 45, for seven years for anyone presently aged 45–50, for five years for anyone presently aged 50–55 and for two years for anyone presently aged over 55. This would mean that a person aged under 45 years old today would work in full-time employment until they are 77 years old. We estimate that if these changes were introduced, it would result in at least a 20% increase to GDP once the transitional arrangements had been worked through. For the US today, this would mean an additional $4 trillion or more annually. If both ideas were in effect today the projected change to GDP would be $5.6 trillion or more, making the US response to the pandemic, which thus far sits at around $5.3 trillion, affordable.

- **Research and Development:** In this book we've listed dozens of areas where research and development is needed, to create the industries and capabilities for the 21st century. But today there's little R&D funding available for those arenas, it is largely left to private investment, and most markets don't have advanced venture capital or private equity markets.

- **Long-Term Programs and Infrastructure Development:** We focus on short-term today. We need to educate the citizenry that long-term planning and projects need to be a core functional element of our economy; that longer-term investments for future generations are needed. But we also need all that dead billionaire wealth and corporate cash reserves that are accumulating in bank accounts to simply be better mobilized.

Political Reform and Policy Debate:

- **Universal Care:** Access to healthcare, housing, education, clothing and internet are basic human rights, according to the United Nations, but access to basic levels of these services is absent in much of the so-called developing world and remains contentious in some developed economies. China successfully eliminated poverty for its poorest 100 million citizens in the space of just eight years, so this can be done.

- **US vs China Economic Competition:** We can expect that the US will not take kindly to the dominance of the Chinese economy over the next decade, with responses ranging from disbelief and accusations of false reporting of economic figures, through to potential conflict in the South China Sea over reclaimed islands planted with Chinese military installations and so forth. The United States will not adjust well to its role as global leader being weakened by an economically dominant China.

- **Ethics and Justice Reform:** To eliminate bias in policing and sentencing, we can expect AI to hand out sentencing and fines. Judges who consistently override

AI-based sentencing will be removed. Increasingly, sentencing will lead to less jail time, more mental health support and community service time geared towards climate mitigation and infrastructure improvement.

- **Immigration Policy:** Due to mass migration of climate refugees, declining birth rates and slowing economic growth, expect competition for skilled resources and global pressure on economic support for non-skilled migrants.

- **Unemployment and UBI:** UBI will become widespread because of increasing unemployment as a result of technology automation.

- **Dealing with Widespread Social Unrest and Protests:** Unemployment will affect crime rates, increasing protests and leading to widespread government dissatisfaction as unemployment grows and is sustained, and as climate impact hits with historically poor government response (e.g., insurance system collapse, displacement, seasonal outages/flooding/air quality issues, etc.). The increasing tendency to clamp down on unrest will reach a crescendo as the numbers of protests dwarf crowd control capabilities, leading to either revolution or genuine reforms.

Luddistan

Capitalism has largely failed, but no new system has emerged

AI, Science and Technology largely rejected

Limits on technology placed into law to keep humans employed and relevant

Technosocialism

Highly automated society

Broad equality and prosperity

Ubiquitous infrastructure for health, education, transport, food and housing

Neo-Feudalism

Enclaves of rich living in walled cities,

Massive inequality reinforced through technology ownership and wealth capture

Failedistan

We responded too late, climate collapse initiated global depression

Hundreds of millions displaced, immigration and resource wars

General Autocratic rule

In the table below, you will find our thoughts of where these policy issues could create forks in the road of social evolution, which pushes us towards certain outcomes.

The responses to specific policy issues are categorized by outcome, where various decisions could lead us to one outcome versus another. Where we designate a strategy or event as rejected, it's generally where we have a worst-case scenario. For example, on AI regulation, one possibility is that as AI takes human jobs we could either tax AI to support UBI, or we could outright ban AI because it's taking human jobs. While saving jobs would be admirable, we can't downsize government without AI, and therefore won't get broader inclusion and equality from mechanisms like UBI.

Event	Year	Category	Rejection	Acceptance	Inaction
First Human on Mars	2028	Space	Ludistan: Too Many Problems at Home	Technosocialism: Multiplanetary Species	
China World's Largest Econom	2030	Economy	Neo-Feudal: This isn't real/How possible?	Technosocialism: Infrastructure/AI Standard	
			Failedistan: This means war!		
Major AI Regulation	2030	AI	Ludistan: Banning AI, Humans First	Technosocialism: AI Ethics for Equality	
Protest Escape Velocity	2030	Politics	Failedistan: Do Nothing, Revolution	Technosocialism: All in this together!	Neo-Feudal: Corporations are People too
Global Flat Tax Rate	2035	Global	Neo-Feudal: Vested Interest Reject	Technosocialism: Citizens before Profits	Failedistan: But market growth?
Human Augmentation	2035	Health	Ludistan: Not the natural order	Technosocialism: Compete with AI	Neo-Feudal: Only for the Rich
Large Scale Techno-unemployr	2035	AI	Ludistan: Ban AI/Robots who take human job	Technosocialism: UBI for the planet	Failedistan: No UBI - who will pay!
				Neo-Feudal: Work harder to keep job	
Longevity Escape Velocity	2040	Health	Ludistan: Not the natural order	Technosocialism: Live Long & Prosper!	Neo-Feudal: Only for the Rich
UBI critical mass	2040	Economy	Failedistan: Do Nothing, Revolution	Technosocialism: UBI for the planet	Neo-Feudal: No UBI - who will pay!
100% Renewable Energy	2045	Energy	Ludistan: Coal is good	Technosocialism: Low Carbon, Free Energy	Neo-Feudal: What does Big Oil say?
Artificial General Intelligence	2045	AI	Ludistan: AI is taking over the world!	Technosocialism: AI Rights Amendment	Neo-Feudal: Pay My AI for the Best Result
Eco-Refugee Boom	2045	Climate	Failedistan: Resource Wars, Closed Borders	Technosocialism: We need a plan!	Neo-Feudal: Who is going to Pay for this?
First Cities Lost to Sea Rise	2045	Global	Failedistan: Can you breathe underwater?	Technosocialism: Let's build resilance	Ludistan: But climate change isn't real...
Global Sustainability Doctrine	2045	Global	Neo-Feudal: But the growth, the markets!	Technosocialism: Live in Harmony with Planet	Failedistan: Business as Usual
National Debt Forgiven	2045	Climate	Failedistan: Who pays what?	Neo-Feudal: Does it have to be climate $$?	Failedistan: Business as Usual
				Technosocialism: Climate Mitigation Fast	
Widespread Crop Failures	2045	Climate	Failedistan: Resource Wars, Famine	Technosocialism: Urban Farming, Lab-Grown	Neo-Feudal: Rich still get the food right?
Western Birth Rates Decline	2055	Economy	Neo-Feudal: Immigrants taking jobs!	Ludistan: We need more babies!	
				Technosocialism: Import people who spend	

What Martian colonization might teach us about economics

Let's embark on one final thought exercise to illustrate how we could potentially reframe the relationship between the human species, capitalism, AI, our planet and our future.

Elon Musk is working hard to make humanity a multi-planetary species. If he and NASA (or through international cooperation) succeed in establishing a Martian colony with 1 million people, what would the Mars economy look like?

The first thing to realize is that capitalism just wouldn't be important, wouldn't be helpful or constructive once humans are on the ground. Mars won't be shipping back resources to Earth, the whole of the martian economy will be focused on one thing—sustainable self-reliance. Musk has said repeatedly that the first goal of a Martian colony would simply be ensuring the colony survives if the Starships from Mars stop resupplying the economy.

Award-winning science fiction author Kim Stanley Robinson wrote an incredibly three-part trilogy on colonizing Mars—*Red Mars*, *Green Mars* and *Blue Mars*. In his books he proposed a radically different type of economy, one based on optimal resource allocation, where people didn't accumulate wealth, they competed to create, live and survive. The assets of the Red-Green-Blue Mars economy were energy abundance and sustainable resource utilization, where colonists prided themselves on producing more energy, food, air, water, and resources than they consumed. The whole colony centered on independence from Earth at the earliest opportunity. Robinson called this ecopoetics in the trilogy, suggesting Martians would have very different motivations from the transnational corporations, long-lived rich elite, and markets of the collapsing Teran economies[7].

If we're going to go to Mars, and we're going to get the Martian colonies to self-sustainability, we're going to need a new doctrine. A doctrine of Sustainable Prosperity, not simply profits and returns.

Capitalists will argue that the only reason Musk is able to think about getting humans to Mars is the success of capitalism itself! It's only because of Musk's extraordinary performance as an entrepreneur creating billions of dollars of wealth, that he can even accomplish this. But there is an argument that in the absence of capitalism, in the absence of the political divide, we might already have humans living on Mars. It's entirely possible we could have made greater technological advancements that the current system has allowed.

The same is true for the future Martian colony. Making a profit, raping the Martian environment of its natural resources, dividing up the spoils for earthly shareholders, will not be a priority in any way shape or form. It would be completely counterproductive for the Martian settlers.

With this framing we might, in fact, create a Martian "central bank" currency that embodies these values. A currency that changes value based on the self-sustainability of the colony. One where your holdings of a currency rise or decline based on how you interact with the economy. If you are putting more resources back into the economy than you take out, you accumulate wealth. If you repair equipment rather than shipping a replacement from Earth, you accumulate wealth. If you are able to live a sustainable existence, and help others, you are contributing to the prosperity of the planet, the viability of the community as a whole.

This is the ultimate in Technosocialism. A system that encourages us to all do well, that pushes our economy to advance for the betterment of the species, and not just for markets and shareholders.

Figure 4: Mars Base Alpha wouldn't operate based on Terran economics.
(Image Credit: Spacex Media Relations).

This would still allow for competition, but competition to optimize resource utilization and optimize the benefits of your business or collective for those around you. This is not Marxist socialism, as ownership simply isn't important in this ecosystem. We all share life on this planet, we all should be able to have a healthy and happy life, living in harmony with the other species that live here.

The thing is, if we could conceivably do this on Mars, we should be able to do this also on Earth. But that would undermine the power of those that sit atop the socio-economic pyramid designed to flow capital and wealth back up to the top of that pyramid. A system designed to create multi-generational wealth for the richest 1 percent, and to advance the interests of their family and their tribe alone, and not for their fellow humans. Ultimately capitalism is a system that is by its nature is designed to create economic division when it is successful, not create a happy, healthy, and sustainable existence. It can create trillions of dollars of wealth for the world's richest billionaires during the worst pandemic in the last 100 years while leaving others homeless, jobless, or worse dead. That is capitalism's greatest flaw because it creates competition that sets us at each other's throats, rather than incentivizing us to cooperate for the good of all. Those that benefit from this system aren't interested in the common good, they're interested in winning.

The conclusion, everything else considered...

There is no such thing as a national climate change policy that fixes the problems for us all—climate change will affect every person on the planet in some way. We're going to need concerted global action to defend humanity against the worst aspects of climate change. Rivalry and competition on a national scale need to be channeled into productive pursuits that can be leveraged by all, not just by those who might make profit from the intellectual property.

Increasingly we're going to need to cooperate across borders. Issues like the coronavirus pandemic and climate change show that we need to get much, much better at working together cross-border to solve the biggest issues.

We've proposed some simple but radical positions to fuel the engines of innovation to solve these problems. Radically more efficient government that dramatically reduces the cost of governance, turning big government into algorithms and automation at a fraction of the cost of the 20th century. We've proposed forgiveness of national debt, but only where that debt is channeled into climate mitigation across the globe. We've proposed a flat global corporate tax rate that incentivizes companies to channel their resources into creating value for all stakeholders, and not simply returning dividends to shareholders.

All of this will create the greatest pool of financial and human resources that the world has ever seen, but it is designed to produce two simple outcomes. An orderly and effective response against AI disruption to employment; and massive incentives for attacking the problems associated with climate change. It allows capitalism to have a much more harmonious platform for the advancement of the species, rather than accentuating divisions between the rich and the poor. There will, of course, be those who argue this is counter to the very essence of capitalism, that competition and not cooperation is key.

Instead of sharing new climate mitigation technologies across borders, today we channel that into making profits for corporations. Profit margins and dividends aren't some economic equivalent of human lives, but today that is how we implicitly measure them. We have the greatest economical and technical capabilities that the world has ever seen, and we're directing that toward creating the world's multi-trillion dollar corporations, Dogecoin, NFTs and billionaires by the bucket loads. One billion people face climate displacement, at least 150 million people have slipped into extreme poverty and homelessness due to the coronavirus pandemic and half of the workers around the world face future displacement by automation over the next 20 years. It is absolutely the time we got our priorities straight—GDP growth and stock market gains are not the equivalent of the future health and happiness of our citizens, and the survivability of all the species that live on our planet.

What's the best way to stop coronavirus or a future pandemic? Vaccinate the world as quickly as possible and ensure adequate treatment resources for all. What's the best way to stop homelessness? Give people homes. What's the best way to stop poverty? Ensure a living wage for all, just as China managed to accomplish in the space of a decade. If you can't do that through jobs, create a Universal Basic income. What's the best way to stop the worst effects of climate change? Mobilize the world's economies to mitigate those effects. What's the best way to mitigate the effects of technology-based unemployment? Ban AI, technology and robots or, alternatively, provide a Universal Basic Income? Which of these choices do you think we should make?

Humanity will be tasked with answering some pretty fundamental, philosophical and practical questions over the next two to three decades. Emerging generations appear to be questioning their parents' cherished values around capitalism, democracy, class, race, and economics. Sometime between now and 2040, for the first time we will likely see a major political shift around the central purpose of the economy, i.e., that the economy's first and primary job is not to make wealth and grow GDP, rather it is to make citizens happy, healthy, and long lived. This will be the first time in history that our economic purpose aligns with the optimal path forward for humanity.

Capitalism built the world's greatest economies, but nonetheless its design is still massively flawed. Capitalism has also allowed the greatest inequality in history; it failed us during the pandemic, and it has failed us when it came to our planet and the thousands of species made extinct by us burning fossil fuels, putting pollutants in the atmosphere, toxins into the environment and dumping plastics in the oceans. We had the technology to be carbon neutral 50 years ago, but we chose short-term profits and returns over the health of our planet. Capitalism is anti-humanity—it works only for the few, not for all.

There are those who believe almost religiously in capitalism's ability to fix these wrongs, but the fact that capitalism created these same problems is not lost on the rest of us. In 10,000 years, the concept that capitalism will be considered the single greatest economic system ever created by humanity is unreasonable. Therefore, we must accept that we can do better, that reform of capitalism is necessary. That we don't have to wait even 100 years for something better.

There's only one way we get to any sort of Utopian and stable future that doesn't divide outcomes dramatically depending on where you live, what economic class you live in, or what your skin colour and gender are. Long-term multi-decade planning and programs—broad and long-term commitments to a very different future, funded as global and national economic development programs. The sort of multi-generational commitment to the future of the planet and species that we simply haven't demonstrated thus far. It needs to be based on a philosophy that transcends market rationale and national boundaries. It needs to be a commitment that brings us together globally, not further divides us based on political systems or economic theory.

The 21st century will bring us events that reshape humanity for the next one hundred generations, technologies and advances that give us the tools and allow us the vision to do extraordinary things. It is an inflection point in the history of the species.

Technosocialism is not a political movement, nor an economic theory. Technosocialism is the inevitability of those two ancient worlds colliding. It is the earth-shattering repercussions of climate change, and the incredible advances of Artificial Intelligence and technology that gives mankind control over its destiny.

It's time for humanity to work towards an optimal state, a sustainable existence on this planet that we share with billions of other creatures, and an optimal future for our grandchildren and our home. Instead of a worldwide descent into insurmountable debt, unemployment, starvation, eco-refugee crises, rolling pandemics and healthcare issues, and the chaotic systems and ideologies that continue to divide us.

It is our future to choose, but only if we abandon the thinking and philosophy that got us here.

Endnotes

1   'Climate change is the defining crisis of our time and it particularly impacts the displaced'; UNHCR's Special Advisor on Climate Action, Andrew Harper, reported by Tim Gaynor, 30 November 2020.

2   Ant Group SEC.gov filings, Fiscal Year 2020 Annual Report: see https://www.sec.gov/Archives/edgar/data/1577552/000110465920082881/a20-6321_46k.pdf.

3   BBC News, "In Your Face: China's All Seeing State", 10 December 2017: https://www.bbc.com/news/av/world-asia-china-42248056.

4   National Institute of Health, Overview of the Shenzhen Emergency Medical Service Call Pattern, Shuk et al.

5   https://www.spglobal.com/_Media/Documents/03651.00_Women_at_Work_Doc.

6   Daley, John, Cassie McGannon, and Leah Ginnivan, Game-changers: *Economic Reform Priorities for Australia*, Grattan Institute, Melbourne, 2012.

7   "Falling into Theory: Simulation, Terraformation, and Eco-Economics in Kim Stanley Robinson's Martian Trilogy", John Hopkins University Press, Volume 43, Number 3, Fall 1997 pp. 773-799.

APPENDIX: THE FOUR POSSIBLE TIMELINES

These are the four potential outcomes for human progress, shown as the dominant timeline in each case.

We've included the potential inflection points that might take us down these roads over the next century. There are other possible outcomes, particularly if some economies become a technosocialist collectivism while others concomitantly become super capitalist highly unequal societies, for example.

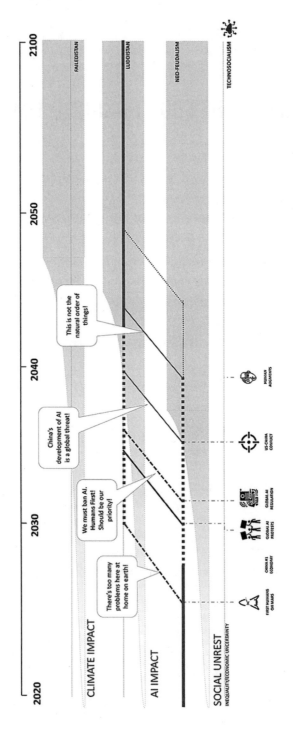

Notes for the Luddistan Timeline

Key inflection points are:

- **General Anti-Science Consensus** — After the failure of governments to adequately control the COVID pandemic and stop climate change, this lends greater support to the voices who distrust science and technology in general. Consensus on policy becomes almost impossible given the division around facts and opinions.

- **Rejection of New Commercial Space Race and Colonization of the Moon/Mars** — deemed irrelevant given humanity's problems with the climate, food scarcity, unemployment, etc.

- **Banning AI due to human employment impact** — Legislation is enacted to prioritize humans over machines, leading to a broad slow down in the impact of automation, and flattening global growth into decline.

- **China targeted** — The EU and US, heavily influenced by public opinion, start to fall behind China economically and technologically due to China's application of AI and automation technologies. This causes military conflict on occasion and certainly economic sanctions for economies employing Chinese technology.

- **Banning Human Augmentation** — Whether it be gene therapy that can flip on or off proteins that eliminate diseases or enhance human performance, or cybernetics and neural interfaces, these technologies will be argued as unethical because they change the natural order of things or because they split humanity into enhanced and natural species.

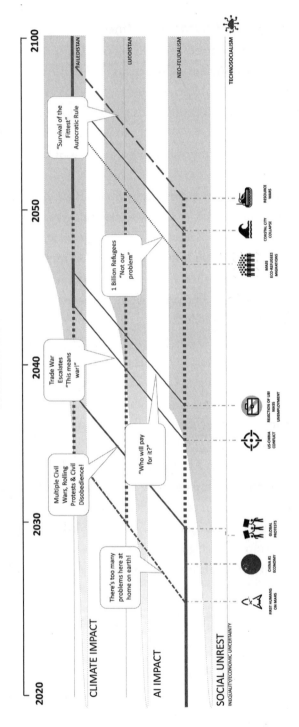

Notes for the Failedistan Timeline

Key inflection points are:

- **General Anti-Science Consensus** — After the failure of governments to adequately control the CoVID pandemic and stop climate change, this lends greater support to the voices who distrust science and technology in general. Consensus on policy becomes almost impossible given the division around facts and opinions.

- **Rejection of New Commercial Space Race and Colonization of the Moon/Mars** — deemed irrelevant given humanity's problems with the climate, food scarcity, unemployment, etc.

- **Mass Global Protests, Governments Overthrown** — Protests span the globe continuously based around various themes. Democracy results in the rapid cycling of short-term majorities that don't give policy enough time to work, resulting in increasingly ineffectual governance.

- **China targeted** — With the US economy slowing and China's economy ramping up, the only thing that will be deemed possible after sanctions fail, is military action against China trade routes with shipping and rail blockades creating military skirmishes, particularly in the South China Sea.

- **Rejection of Universal Basic Income** — While large sections of the population plunge into unemployment, consensus is against Universal Basic Income because of arguments over who will fund UBI and how it will be funded. Corporations successfully lobby US Congress to agree not to increase taxes on corporations, essentially killing off UBI.

- **Global Eco-Refugee Crisis** — With coastal cities inundated, 100 year floods annually, drought, wildfires and hurricanes impacting cities, along with large crop failures and government failures, the 1 billion eco-refugees looking for a home are turned away at every corner, creating the greatest humanitarian disaster since the bubonic plague.

- **Survival of the Fittest** — With governments increasingly ineffective, autocrats promise to fill the vacuum but increasingly reducing personal security and safety. The rich remain protected by governments courting their support, but for most of the population the lack of global consensus results in poverty, homelessness, declining health and high crime impact.

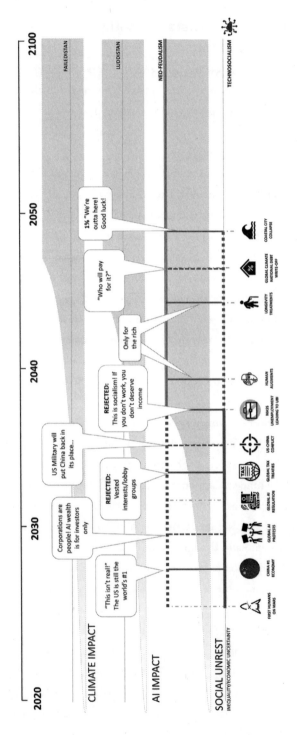

Notes for the Neo-Feudalism Timeline

Key inflection points are:

- **East vs Western Battlelines** — When China becomes the world's largest economy, the West largely rejects the premise, finding ways to reclassify China's economic activity, and increasingly sanction and blockade China. The Belt and Road Initiative faces severe operational difficulties due to geo-political impact, China's growth slows, but still outpaces the US.

- **Corporations are People Too** — As global protests breakout, corporations spend billions on influencing policy to continue to benefit their shareholders and employees. Protests become very focused on the control and power of corporations but are ineffectual at getting real change in most democracies.

- **Rejection of Global Tax Treaty** — Again corporations fight to retain their right to move funds around the planet to minimize taxation. This means global tax treaties fall apart and domiciling for tax becomes a competitive feature of economies trying to attract large corporate head offices. Taxation revenue from corporations continues to decline, putting more and more pressure on lower-to-middle class tax brackets, supercharging inequality.

- **China targeted** — With the US economy slowing and China's economy ramping up, the only thing that will be deemed possible after sanctions fail is military action against China's trade routes with shipping and rail blockades leading to military skirmishes, particularly in the South China Sea.

- **Rejection of Universal Basic Income** — This is socialism!! While large sections of the population plunge into unemployment, consensus is against Universal Basic Income because of arguments over who will arguments over who will fund UBI and how it will be funded. Corporations successfully lobby US Congress to agree not to increase taxes on corporations essentially killing off UBI.

- **Longevity Treatments** — Only the richest 1% will have access to such technology as the high cost ends up creating a small market where costs don't materially decrease because of scale benefits. Thus, treatments remaining accessible only to the richest.

- **Rejection of National Debt Write-Off** — The write-off of national debt to dedicate to climate mitigation, techno-unemployment, etc is rejected as an unlevel playing field, with some governments benefiting more than others. Slows climate response, adaptation to automation and AI standards, improvements in science and technology more broadly.

- **We're outta here** — The Richest 1% increasingly threatened by violent mobs with torches and pitchforks move to geographies willing to protect them and offer them advantages on the fringes of ethical economics.

- **Neo-Feudalism** — The Techno Feudal Lords will rule over humanity living for hundreds of years with some global corporations holding as much power as entire governments and nations, with more wealth too. They set the parameters of global rule through the ability to manage checkbook consensus.

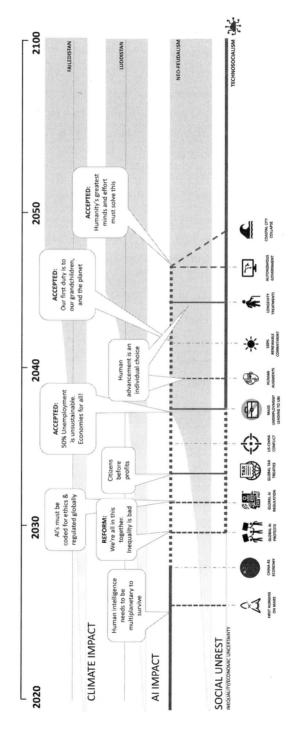

Notes for the Technosocialism Timeline

Key inflection points are:

- **Multi-Planetary** — Humanity embraces being a mutli-planetary species. This is seen as inspirational and encouraging massive technological progress, while also guaranteeing humanity's future in the face of climate change or potential future extinction level events.

- **Global Reform Movement** — starting with the populist movement, humanity embarks on multiple decade long reform movements around inequality, universal basic care (Globally), the application of technology on healthcare, climate mitigation, etc.

- **Ethical AI & Regulation** — As AI's impact grows, AI creators are bound by global regulation and standards around AI training models and matrixes. Human ethics are encoded at a global level and built into AIs that are deployed in government or broad services automation.

- **Citizens before Profits** — As techno-unemployment continues to reshape our nations, corporations that are responsible for eliminating millions of human workers are co-opted into new UBI initiatives and human retraining and reemployment programs. Including jobs specifically targeting climate mitigation and resilience, factory food production, and universal services — education, healthcare, housing.

- **Universal Basic Income** — UBI is accepted as a global standard for countries facing 30% unemployment or more due to technology and climate change.

- **Meta-Humans** — The era of human enhancement and accelerated evolution starts. While seen as a personal choice, the advantages of basic augmentation will be significant. This will result in two classes of humans overtime, but with natural unenhanced humans protected at law and seen as a valuable part of the genetic baseline humanity.

- **100% Renewable** — Commitment globally to zero fossil fuel usage by 2050 is expected, maybe before. Not just because it is significantly cheaper, but because it also reduces CO^2 emissions and reduces deaths for pollution, etc. Energy effectively becomes free for all.

- **The Greatest Minds** — Collectively the scientists, technologists, corporations, economists and politicians all come together across borders to attack the global problem of climate change. This results in the greatest global cooperation technically and organizationally that we've ever seen. Humanity will dedicate the next 50–100 years to prioritize our human habitat, the Earth, and ensuring that the 6th great extinction will be slowed or even repaired. Sustainability and renewability become sacred rules that humanity lives by. This pushes us to mine asteroids and other bodies for resources while protecting the natural resources on earth.

- **Technosocialism** — Humanity thrives as technology allows us all individually and collectively to reach our optimal potential as a species. We live in harmony with the planet and each other. Poverty is eliminated. Human longevity is increased exponentially. Our intelligence survives and thrives. We don't allow the abstraction of money or national boundaries any longer to hold us back.

ABOUT THE AUTHORS

Brett King

Brett King is a world-renowned entrepreneur, futurist, speaker, international bestselling author, and media personality. China's President Xi Jinping cited his book *Augmented: Life in the Smart Lane* on the topic of Artificial Intelligence in his 2018 national address; the same book that was listed as a Top 10 non-fiction book in North America. In 2019 his book *Bank 4.0* was awarded the Top Book by a Foreign Author in Russia for that year.

In 2020 King was inducted into the Fintech Hall of Fame by CB Insights. In 2015 he was shortlisted for the Advance Global Australian of the Year Award. His books have been released in over a dozen languages. *Banking Exchange* magazine dubbed him the "King of Disruptors", while *The Australian* newspaper in Australia called him the "Godfather of Fintech".

Brett has spoken in more than 50 countries, at TED, *Wired*, Techsauce, Singularity University, Web Summit, *The Economist*, IBM's World of Watson, CES, SIBOS and many more. He has appeared as a commentator on CNBC, BBC, ABC, Fox, and Bloomberg. He has spoken live to virtual audiences of over 42 million people.

He advised the Obama administration on Fintech strategy, and today advises regulators, lawmakers and boardrooms around the world on digital transformation and future readiness. King founded Breaking Banks and The Futurists podcasts (distribution to 180 countries, 9 million annual listeners). King founded the world's first mobile challenger bank—Moven —in 2011, raising $47m to-date and sits on the board of the largest US non-profit focused on financial inclusion, the Financial Health Network.

Richard Petty

Dr Richard Petty is a government policy advisor, entrepreneur and awarded academic. Based in Hong Kong, Richard has lived and worked in Greater China since the 1990's. He is a past Chairman of the Australian Chamber of Commerce Hong Kong & Macau and CPA Australia, and is founding Vice Chairman, Hong Kong ASEAN Economic Cooperation Foundation. Richard sits on the board of listed companies in the United States and Australia, has advised on projects with an aggregate economic impact in the hundreds of billions of dollars, and has led reviews of several economies and studies on economic competitiveness in several countries.

A former board member of International Federation of Accountants (IFAC), for many years Richard served as a member of the B20—the official G20 dialogue forum with the global business community—on the Financing Growth and Infrastructure Taskforce.

Richard has been professor or visiting professor at several business schools and delivered lectures and addresses at many universities including Stanford Graduate School of Business, Macquarie Graduate School of Management, Cambridge University, University of New South Wales, Kellogg-HKUST, The University of Sydney, Emory University, and The University of Hong Kong. He has presented in more than 20 countries as an invited speaker for organizations including the OECD, Pacific Basin Economic Council, and Hong Kong Federation of Industries, and provided commentary on CNBC, Bloomberg, BBC, ABC, *The Wall Street Journal*, *The South China Morning Post*, and many others.

Richard has authored or co-authored more than 100 academic and professional works including several books, and several of his journal articles are among the top cited in their fields. He has served on the editorial boards of numerous academic journals. Richard holds several degrees, including a PhD. He was awarded the university medal for his undergraduate studies. He is a Fellow of Chartered Accountants Australia and New Zealand, CPA Australia, and the Australian Institute of Company Directors.